Argument-Driven Inquiry
in
LIFE SCIENCE

LAB INVESTIGATIONS
for GRADES 6–8

Argument-Driven Inquiry
in
LIFE SCIENCE

LAB INVESTIGATIONS
for GRADES 6–8

Patrick J. Enderle, Ruth Bickel, Leeanne Gleim, Ellen Granger,
Jonathon Grooms, Melanie Hester, Ashley Murphy, Victor Sampson,
and Sherry A. Southerland

NSTApress
National Science Teachers Association
Arlington, Virginia

National Science Teachers Association

Claire Reinburg, Director
Wendy Rubin, Managing Editor
Andrew Cooke, Senior Editor
Amanda O'Brien, Associate Editor
Donna Yudkin, Book Acquisitions Coordinator

Art and Design
Will Thomas Jr., Director

Printing and Production
Catherine Lorrain, Director

National Science Teachers Association
David L. Evans, Executive Director
David Beacom, Publisher

1840 Wilson Blvd., Arlington, VA 22201
www.nsta.org/store
For customer service inquiries, please call 800-277-5300.

NSTA is committed to publishing material that promotes the best in inquiry-based science education. However, conditions of actual use may vary, and the safety procedures and practices described in this book are intended to serve only as a guide. Additional precautionary measures may be required. NSTA and the authors do not warrant or represent that the procedures and practices in this book meet any safety code or standard of federal, state, or local regulations. NSTA and the authors disclaim any liability for personal injury or damage to property arising out of or relating to the use of this book, including any of the recommendations, instructions, or materials contained therein.

Library of Congress Cataloging-in-Publication Data
    Enderle, Patrick.
      Argument-driven inquiry in life science : lab investigations for grades 6-8 / Patrick J. Enderle [and 8 others].
        pages cm
      Includes bibliographical references and index.
      ISBN 978-1-938946-24-0 (print) -- ISBN 978-1-941316-73-3 (e-book) 1. Biology--Methodology--Study and teaching (Middle school) 2. Biology--Experiments. 3. Experimental design--Study and teaching (Middle school). I. Title.
      QH324.E483 2015
      570.78--dc23
                     2015013671

    Cataloging-in-Publication Data for the e-book are also available from the Library of Congress.
    e-LCCN: 2015021523

CONTENTS

SECTION 1
Using Argument-Driven Inquiry

SECTION 2—Life Sciences Core Idea 1
From Molecules to Organisms: Structures and Processes

INTRODUCTION LABS

APPLICATION LABS

SECTION 3—Life Sciences Core Idea 2

Ecosystems: Interactions, Energy, and Dynamics

INTRODUCTION LABS

APPLICATION LABS

SECTION 4—Life Sciences Core Idea 3
Heredity: Inheritance and Variation in Traits

INTRODUCTION LABS

APPLICATION LAB

SECTION 5—Life Sciences Core Idea 4
Biological Evolution: Unity and Diversity

INTRODUCTION LAB

APPLICATION LABS

SECTION 6—Appendixes

PREFACE

There is a push to change the way science is taught in the United States, called for by a different idea of what it means to know, understand, and be able to do in science. As described in *A Framework for K–12 Science Education* (National Research Council [NRC] 2012) and the *Next Generation Science Standards* (NGSS Lead States 2013), science education should be structured to emphasize ideas *and* practices to

> ensure that by the end of 12th grade, *all* students have some appreciation of the beauty and wonder of science; possess sufficient knowledge of science and engineering to engage in public discussions on related issues; are careful consumers of scientific and technological information related to their everyday lives; are able to continue to learn about science outside school; and have the skills to enter careers of their choice, including (but not limited to) careers in science, engineering, and technology. (p. 1)

Instead of teaching with the goal of helping students learn facts and concepts, science teachers are now charged with helping their students become *proficient* in science by the time they graduate from high school. To allow for this proficiency, the NRC (2012) suggests that students need to understand four core ideas in the life sciences,[1] be aware of seven crosscutting concepts that span across the various disciplines of science, and learn how to participate in eight fundamental scientific practices in order to be considered proficient in science. These important practices, crosscutting concepts, and core ideas are summarized in Figure 1 (p. xii).

As described by the NRC (2012), new instructional approaches are needed to assist students in developing these proficiencies. This book provides 20 lab activities designed using an innovative approach to lab instruction called argument-driven inquiry (ADI). This approach and the labs based on it are aligned with the content, crosscutting concepts, and scientific practices outlined in Figure 1. Because the ADI model calls for students to give presentations to their peers, respond to questions, and then write, evaluate, and revise reports as part of each lab, the lab activities described in this book will also enable students to develop the disciplinary-based literacy skills outlined in the *Common Core State Standards* for English language arts (NGAC and CCSSO 2010). Use of these labs, as a result, can help teachers align their instruction with current recommendations for making life science more meaningful for students and more effective for teachers.

---

1 Throughout this book, we use the term *life sciences* when referring to the core ideas of the *Framework* (in this context the term refers to a broad collection of scientific fields), but we use the term *life science* when referring to courses at the middle school level (as in the title of the book).

FIGURE 1

The three dimensions of the framework for the *NGSS*

Scientific Practices	Crosscutting Concepts
1. Asking questions and defining problems	1. Patterns
2. Developing and using models	2. Cause and effect: Mechanism and explanation
3. Planning and carrying out investigations	3. Scale, proportion, and quantity
4. Analyzing and interpreting data	4. Systems and system models
5. Using mathematics and computational thinking	5. Energy and matter: Flows, cycles, and conservation
6. Constructing explanations and designing solutions	6. Structure and function
7. Engaging in argument from evidence	7. Stability and change
8. Obtaining, evaluating, and communicating information	

Life Sciences Core Ideas

- LS1: From molecules to organisms: Structures and processes
- LS2: Ecosystems: Interactions, energy, and dynamics
- LS3: Heredity: Inheritance and variation of traits
- LS4: Biological evolution: Unity and diversity

Source: Adapted from NRC 2012, p. 3.

References

National Governors Association Center for Best Practices and Council of Chief State School Officers (NGAC and CCSSO). 2010. *Common core state standards.* Washington, DC: NGAC and CCSSO.

National Research Council (NRC). 2012. *A framework for K–12 science education: Practices, crosscutting concepts, and core ideas.* Washington, DC: National Academies Press.

NGSS Lead States. 2013. *Next Generation Science Standards: For states, by states.* Washington, DC: National Academies Press. *www.nextgenscience.org/next-generation-science-standards.*

ACKNOWLEDGMENTS

The development of this book was supported by the Institute of Education Sciences, U.S. Department of Education, through grant R305A100909 to Florida State University. The opinions expressed are those of the authors and do not represent the views of the institute or the U.S. Department of Education.

ABOUT THE AUTHORS

Patrick J. Enderle is a research faculty member in the Center for Education Research in Mathematics, Engineering, and Science (CERMES) at The University of Texas at Austin (UT-Austin). He received his BS and MS in molecular biology from East Carolina University. Patrick then spent some time as a high school biology teacher and several years as a visiting professor in the Department of Biology at East Carolina University. He then attended Florida State University (FSU), where he graduated with a PhD in science education. His research interests include argumentation in the science classroom, science teacher professional development, and enhancing undergraduate science education. To learn more about his work in science education, go to *http://patrickenderle.weebly.com*.

Ruth Bickel has been a teacher at FSU Schools for several years, supporting student learning in a variety of disciplines. She was originally a social studies teacher before taking an interest in teaching science. She has taught middle school Earth and space science and life science for several years. She has also taught a high school–level forensics course over the past few years. Ruth was responsible for writing and piloting many of the lab investigations included in this book.

Leeanne Gleim received a BA in elementary education from the University of Southern Indiana and an MS in science education from FSU. While at FSU, she worked as a research assistant for Victor Sampson (see his biography later in this section). After graduating, she taught biology and honors biology at FSU Schools, where she participated in the development of the argument-driven inquiry model. Leeanne was also responsible for writing and piloting many of the lab investigations included in this book.

Ellen Granger is the director of the Office of Science Teaching Activities and co-director of FSU-Teach, a collaborative math and science teacher preparation program between the College of Arts and Sciences and the College of Education at FSU. She earned her doctorate in neuroscience from FSU. She is a practicing scientist and science educator and has worked in teacher professional development for almost 20 years. In November 2013, she was named a Fellow of the American Association for the Advancement of Science for "distinguished contribution, service and leadership in advancing knowledge and classroom practices in science education."

Jonathon Grooms received a BS in secondary science and mathematics teaching with a focus in chemistry and physics from FSU. Upon graduation, Jonathon joined FSU's Office of Science Teaching Activities, where he directed the physical science outreach program Science on the Move. He entered graduate school at FSU and earned a PhD in science education. He now serves as a research scientist in CERMES (Center for Education Research in Mathematics, Engineering, and Science) at FSU. To learn more about his work in science education, go to *www.jgrooms.com*.

Melanie Hester has a BS in biological sciences with minors in chemistry and classical civilizations from FSU and an MS in secondary science education from FSU. She has been teaching for more than 20 years, with the last 13 at the FSU School in Tallahassee. Melanie was a Lockheed Martin fellow and a Woodrow Wilson fellow and received a Teacher of the Year award in 2007. She frequently gives presentations about innovative approaches to teaching at conferences and works with preservice teachers. Melanie was responsible for writing and piloting many of the lab investigations included in this book.

Ashley Murphy attended FSU and earned a BS with dual majors in biology and secondary science education. Ashley spent some time as a middle school biology and science teacher before entering graduate school at UT-Austin, where she is currently working toward a PhD in STEM (science, technology, engineering, and mathematics) education. Her research interests include argumentation in elementary and middle school classrooms. As an educator, she frequently employed argumentation as a means to enhance student understanding of concepts and science literacy.

Victor Sampson is an associate professor of science education and the director of CERMES at UT-Austin. He received a BA in zoology from the University of Washington, an MIT from Seattle University, and a PhD in curriculum and instruction with a specialization in science education from Arizona State University. Victor taught high school biology and chemistry for nine years before taking a position at FSU and then moving to UT-Austin. He specializes in argumentation in science education, teacher learning, and assessment. To learn more about his work in science education, go to *www.vicsampson.com.*

Sherry A. Southerland is a professor at FSU and the co-director of FSU-Teach. She received a BS and an MS in biology from Auburn University and a PhD in curriculum and instruction from Louisiana State University, with a specialization in science education and evolutionary biology. Sherry has worked as a teacher educator, biology instructor, high school science teacher, field biologist, and forensic chemist. Her research interests include understanding the influence of culture and emotions on learning—specifically evolution education and teacher education—and understanding how to better support teachers in shaping the way they approach science teaching and learning.

INTRODUCTION

The Importance of Helping Students Become Proficient in Science

The new aim of science education in the United States is for all students to become proficient in science by the time they finish high school. It is essential to recognize that science proficiency involves more than an understanding of important concepts, it also involves being able to *do* science. *Science proficiency*, as defined by Duschl, Schweingruber, and Shouse (2007), consists of four interrelated aspects. First, it requires an individual to know important scientific explanations about the natural world, to be able to use these explanations to solve problems, and to be able to understand new explanations when they are introduced to the individual. Second, it requires an individual to be able to generate and evaluate scientific explanations and scientific arguments. Third, it requires an individual to understand the nature of scientific knowledge and how scientific knowledge develops over time. Finally, and perhaps most important, an individual who is proficient in science should be able to participate in scientific practices (such as designing and carrying out investigations and arguing from evidence) and communicate in a manner that is consistent with the norms of the scientific community.

In the past decade, however, the importance of learning how to participate in scientific practices has not been acknowledged in the standards of many states. Many states have also attempted to make their science standards "more rigorous" by adding more content to them or lowering the grade level at which content is introduced rather than by emphasizing depth of understanding of core ideas and crosscutting concepts, as described by the National Research Council (NRC) in *A Framework for K–12 Science Education* (NRC 2012). The result of the increased number of science standards and the pressure to "cover" them to prepare students for high-stakes tests that target facts and definitions is that teachers have "alter[ed] their methods of instruction to conform to the assessment" (Owens 2009, p. 50). The unintended consequences of this change has been a focus on content (learning "facts") rather than on developing scientific habits of mind or participating in the practices of science. Teachers must move through the curriculum quickly before the administration of the tests, forcing them to cover many topics in a shallow fashion rather than to delve into them deeply to foster understanding.

Despite this high-stakes accountability for science learning, students do not seem to be gaining proficiency in science. According to *The Nation's Report Card: Science 2009* (National Center for Education Statistics 2011), only 21% of all 12th-grade students who took the National Assessment of Educational Progress in science scored at the proficient level. The performance of U.S. students on international assessments is even bleaker, as indicated by their scores on the science portion of the Programme for International Student Assessment (PISA). PISA is an international study that was launched by the Organisation for Economic Co-operation and Development (OECD) in 1997, with the goal of assessing education systems worldwide; more than 70 countries have participated in the study. The test is designed to assess reading, math, and science achievement and is given every three years. The mean score for students in the United States on the science portion of the PISA in 2012 is below the international mean (500), and there has been no significant change in the U.S. mean score since 2000; in fact, the U.S. mean score in 2012 is slightly less than it was in 2000 (OECD 2012; see Table 1). Students in countries such China, Korea, Japan, and Finland score significantly higher than students in the United States. These results suggest that U.S. students are not becoming proficient in science, even though teachers are covering a great deal of material and being held accountable for it.

TABLE 1

PISA scientific literacy performance for U.S. students

Year	U.S. mean score*	U.S. rank/Number of countries assessed	Top three performers
2000	499	14/27	Korea (552) Japan (550) Finland (538)
2003	491	22/41	Finland (548) Japan (548) Hong Kong–China (539)
2006	489	29/57	Finland (563) Hong Kong–China (542) Canada (534)
2009	499	15/43	Japan (552) Korea (550) Hong Kong–China (541)
2012	497	36/65	Shanghai–China (580) Hong Kong–China (555) Singapore (551)

*The mean score of the PISA is 500 across all years.
Source: OECD 2012.

Additional evidence of the consequences of emphasizing breadth over depth comes from empirical research in science education supporting the notion that broad, shallow coverage neglects the practices of science and hinders the development of science proficiency (Duschl, Schweingruber, and Shouse 2007; NRC 2005, 2008). As noted in the *Framework* (NRC 2012),

> K–12 science education in the United States fails to [promote the development of science proficiency], in part because it is not organized systematically across multiple years of school, emphasizes discrete facts with a focus on breadth over depth, and does not provide students with engaging opportunities to experience how science is actually done. (p. 1)

Based on their review of the available literature, the NRC recommends that science teachers delve more deeply into core ideas to help their students develop improved understanding and retention of science content. The NRC also calls for students to be given more experience participating in the practices of science, with the goal of enabling students to better engage in public discussions about scientific issues related to their everyday lives, to be consumers of scientific information, and to have the skills and abilities needed to enter science or science-related careers. We think the school science laboratory is the perfect place to focus on core ideas and engage students in the practices of science and, as a result, help them develop the knowledge and abilities needed to be proficient in science.

How School Science Laboratories Can Help Foster the Development of Science Proficiency

Investigators have shown that lab activities have a standard format in U.S. secondary-school classrooms (Hofstein and Lunetta 2004; NRC 2005). (We use the NRC's definition of a school science lab activity, which is "an opportunity for students to interact directly with the material world using the tools, data collection techniques, models, and theories of science" [NRC 2005, p. 3].) This format begins with the teacher introducing students to a concept through direct instruction, usually a lecture and/or reading. Next, students complete a confirmatory laboratory activity, usually following a "cookbook recipe" in which the teacher provides a step-by-step procedure to follow and a data table to fill out. Finally, students are asked to answer a set of focused analysis questions to ensure that the lab has illustrated, confirmed, or otherwise verified the targeted concept(s). This type of approach does little to promote science proficiency because it often fails to help students think critically about the concepts, engage in important scientific practices (such as designing an investigation, constructing explanations, or arguing from evidence), or develop scientific habits of mind (Duschl, Schweingruber, and Shouse 2007; NRC 2005). Further, this approach does not perceptibly improve communication skills.

Changing the focus of lab instruction can help address these challenges. To implement such a change, teachers will have to emphasize "how we know" in the life sciences (i.e., how new knowledge is generated and validated) equally with "what we know" about life on Earth (i.e., the theories, laws, and unifying concepts). Because it is an essential practice of science, the NRC calls for *argumentation* (defined as proposing, supporting, and evaluating claims on the basis of reason) to play a more central role in the teaching and learning of science. The NRC (2012) provides a good description of the role argumentation plays in science:

> Scientists and engineers use evidence-based argumentation to make the case for their ideas, whether involving new theories or designs, novel ways of collecting data, or interpretations of evidence. They and their peers then attempt to identify weaknesses and limitations in the argument, with the ultimate goal of refining and improving the explanation or design. (p. 46)

This means that the focus of teaching will have to shift more to scientific abilities and habits of mind so that students can learn to construct and support scientific knowledge claims through argument (NRC 2012). Students will also have to learn to evaluate the claims or arguments made by others.

A part of this change in instructional focus will need to be a change in the nature of lab activities (NRC 2102). Students will need to have more experiences engaging in scientific practices so that lab activities can become more authentic. This is a major shift away from labs driven by prescribed worksheets and data tables to be completed. These activities will have to be thoughtfully constructed so as to be educative and help students develop the required knowledge, skills, abilities, and habits of mind. This type of instruction will require that students receive feedback and learn from their mistakes; hence, teachers will need to develop more strategies to help students learn from their mistakes.

The argument-driven inquiry (ADI) instructional model (Sampson and Gleim 2009; Sampson, Grooms, and Walker 2009, 2011) was designed as a way to make lab activities more authentic and educative for students and thus help teachers promote and support the development of science proficiency. This instructional model reflects research about how people learn science (NRC 1999) and is also based on what is known about how to engage students in argumentation and other important scientific practices (Berland and Reiser 2009; Erduran and Jimenez-Aleixandre 2008; McNeill and Krajcik 2008; Osborne, Erduran, and Simon 2004; Sampson and Clark 2008).

Organization of This Book

The remainder of this book is divided into six sections. Section 1 begins with two chapters describing the ADI instructional model and the development and components of the ADI lab investigations. Sections 2–5 contain the lab investigations, including notes for the teacher, student handouts, and checkout questions. Section 6 contains four appendixes with standards alignment matrixes, timeline and proposal options for the investigations, and a form for assessing the investigation reports.

Safety Practices in the Science Laboratory

It is important for science teachers to make hands-on and inquiry-based lab activities as safe as possible for students. Teachers therefore need to have proper engineering controls (e.g., fume hoods, ventilation, fire extinguisher, eye wash/shower), standard operating safety procedures (e.g., chemical hygiene plan, board of education/school safety policies), and appropriate personal protective equipment (sanitized indirectly vented chemical-splash goggles, gloves, aprons, etc.) in the classroom, laboratory, or field during all hands-on activities. Teachers also need to adopt legal safety standards and enforce them inside the classroom. Finally, teachers must review and comply with all safety polices and chemical storage and disposal protocols that have been established by their school district or school.

Throughout this book, safety precautions are provided for each investigation. Teachers should follow these safety precautions to provide a safer learning experience for students. The safety precautions associated with each activity are based, in part, on the use of the recommended materials and instructions, legal safety compliance standards, and current better professional safety practices. Selection of alternative materials or procedures for these activities may jeopardize the level of safety and therefore is at the user's own risk. We also recommend that students, before working in the laboratory for the first time, review the National Science Teacher Association's safety acknowledgment form in the document *Safety in the Science Classroom, Laboratory, or Field Sites* under the direction of the teacher. This document is available online at *www.nsta.org/docs/SafetyInTheScienceClassroomLabAndField.pdf.* The students and their parents or guardians should then sign this document to acknowledge that they understand the safety procedures that must be followed during a lab activity. Additional safety compliance resources can be found on the NSTA safety portal at *www.nsta.org/safety.*

References

Berland, L., and B. Reiser. 2009. Making sense of argumentation and explanation. *Science Education* 93 (1): 26–55.

Duschl, R. A., H. A. Schweingruber, and A. W. Shouse, eds. 2007. *Taking science to school: Learning and teaching science in grades K–8*. Washington, DC: National Academies Press.

Erduran, S., and M. Jimenez-Aleixandre, eds. 2008. *Argumentation in science education: Perspectives from classroom-based research*. Dordrecht, The Netherlands: Springer.

Hofstein, A., and V. Lunetta. 2004. The laboratory in science education: Foundations for the twenty-first century. *Science Education* 88: 28–54.

McNeill, K., and J. Krajcik. 2008. Assessing middle school students' content knowledge and reasoning through written scientific explanations. In *Assessing science learning: Perspectives from research and practice*, eds. J. Coffey, R. Douglas, and C. Stearns, 101–116. Arlington, VA: NSTA Press.

National Center for Education Statistics. 2011. *The nation's report card: Science 2009*. Washington, DC: U.S. Department of Education.

National Research Council (NRC). 1999. *How people learn: Brain, mind, experience, and school*. Washington, DC: National Academies Press.

National Research Council (NRC). 2005. *America's lab report: Investigations in high school science*. Washington, DC: National Academies Press.

National Research Council (NRC). 2008. *Ready, set, science: Putting research to work in K–8 science classrooms*. Washington, DC: National Academies Press.

National Research Council (NRC). 2012. *A framework for K–12 science education: Practices, crosscutting concepts, and core ideas*. Washington, DC: National Academies Press.

Organisation for Economic Co-operation and Development (OECD). 2012. OECD Programme for International Student Assessment. *www.oecd.org/pisa*.

Osborne, J., S. Erduran, and S. Simon. 2004. Enhancing the quality of argumentation in science classrooms. *Journal of Research in Science Teaching* 41 (10): 994–1020.

Owens, T. 2009. Improving science achievement through changes in education policy. *Science Educator* 18 (2): 49–55.

Sampson, V., and D. Clark. 2008. Assessment of the ways students generate arguments in science education: Current perspectives and recommendations for future directions. *Science Education* 92 (3): 447–472.

Sampson, V., and L. Gleim. 2009. Argument-Driven Inquiry to promote the understanding of important concepts and practices in biology. *American Biology Teacher* 71 (8): 471–477.

Sampson, V., J. Grooms, and J. Walker. 2009. Argument-Driven Inquiry: A way to promote learning during laboratory activities. *The Science Teacher* 76 (7): 42–47.

Sampson, V., J. Grooms, and J. Walker. 2011. Argument-Driven Inquiry as a way to help students learn how to participate in scientific argumentation and craft written arguments: An exploratory study. *Science Education* 95 (2): 217–257.

SECTION 1
Using Argument-Driven Inquiry

CHAPTER 1
Argument-Driven Inquiry

Stages of Argument-Driven Inquiry

The argument-driven inquiry (ADI) instructional model consists of eight stages (Figure 2) designed to increase science proficiency by engaging students in the practices of science. Thus, the laboratory investigation becomes more "authentic," including a chance for students to interpret findings from student-designed investigations through creating an explanation for them and arguing this explanation with their classmates. Feedback and guidance are built into the model in such a way that all four areas of science proficiency (as described in the "Introduction" section of this book) should improve over the school year in classes that *regularly* engage in ADI. Each of the eight stages is described in detail in this chapter.

FIGURE 2

Stages of the argument-driven inquiry instructional model

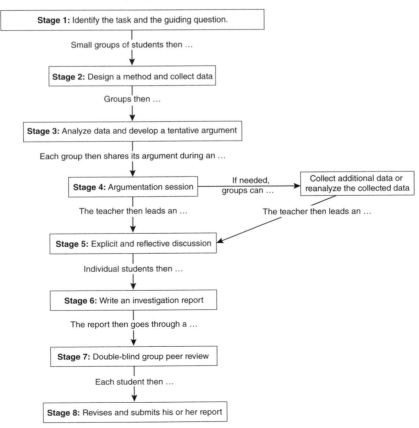

The ADI instructional model encourages student interactions at multiple stages that require students to participate in complex literacy and discourse practices. This model provides teachers with a method for addressing standards identified in the *Common Core State Standards* in English language arts (*CCSS ELA*) and mathematics (*CCSS Mathematics*) (NGAC and CCSSO 2012). However, students in many classrooms may not be fully equipped or have familiarity with the specific kinds of literacy and discourse practices present in the ADI instructional model. Therefore, in the following sections, once the activities of a particular stage are described, we identify some general strategies you can use to help students with literacy challenges. These ideas have been developed during the original research project and through our continued work with multiple groups of teachers around the country. These suggestions are general and may not necessarily be appropriate for your particular group of students and the accommodations they require. We strongly encourage you to seek out more specific assistance for these challenges from your colleagues at your school, including English teachers and literacy coaches.

Stage 1: Identification of the Task and the Guiding Question; "Tool Talk"

In the ADI instructional model each lab activity begins with the teacher identifying a phenomenon to investigate and offering a guiding question for the students to answer. The goal of the teacher at this stage of the model is to capture the students' interest and provide them with a reason to complete the investigation. To aid in this, the teacher should provide each student with a copy of the Lab Handout. The handout includes a brief introduction that describes a puzzling phenomenon or a problem to solve and provides a guiding question to answer. This handout also includes information about the medium students will use to present their argument (e.g., a whiteboard), some helpful tips on how to get started, and the criteria that will be used to judge argument quality (e.g., the sufficiency of the explanation and the quality of the evidence). At the middle school level, it is important for the teacher to carefully clarify each section of the handout and the investigation's expectations out loud and to answer questions on each section.

For students who experience challenges with reading scientific and technical texts, we recommend that you use reading comprehension strategies to help them during this stage. Such strategies can include "decoding" processes, which typically involve students making notations on a document to highlight certain elements. A simple decoding strategy could include having students underline the text that describes concepts related to the activity in the lab investigation, put a box around information that can help them design their investigation, and circle unfamiliar terms in the text. They can use the left margin space for writing notes and the right margin space for writing questions. This decoding process can also be done while students are reading the text out loud.

It is also important for the teacher to hold a "tool talk" during this stage, taking a few minutes to explain how to use specific lab equipment, specific indicators, computer simulations, or software to analyze data. A tool talk is needed because students are often unfamiliar with lab equipment; even if they are familiar with the equipment, they will often use it incorrectly or in an unsafe manner. A tool talk can also be productive during this stage because students often find it difficult to design a method to collect the data needed to answer the guiding question (the task of stage 2) when they do not understand how to use the available materials. The teacher should also review specific safety protocols and precautions as part of the tool talk.

Once all the students understand the goal of the activity and how to use the available materials, the teacher should divide the students into small groups (we recommend three students per group to allow for optimal engagement) and move on to the second stage of the ADI model.

Stage 2: Designing a Method and Collecting Data

In this stage of the ADI model, small groups of students work together to (1) develop a method that they can use to gather the data needed to answer the guiding question and then (2) carry out the method. This stage also includes a review of relevant safety compliance procedures (personal protective equipment, etc.). How students complete this stage depends on the nature of the investigation. Some investigations call for groups to answer the guiding question by designing a controlled experiment, whereas others require students to analyze an existing data set (e.g., a database, information sheets). To assist students with the process of designing their method, the teacher can have students complete an investigation proposal. These proposals guide students through the process of developing a method by encouraging them to think about what type of data they will need to collect, how to collect it, and how to analyze it. We have included three different investigation proposals in Appendix 3 (p. 361): Investigation Proposal A or Investigation Proposal B can be used when students need to design a method to test alternative explanations or claims; Investigation Proposal C can be used when students need to collect systematic observations and do not need to design a method to test alternative explanations or claims. The proposals also provide scaffolding to help structure and support students' discussion. Further assistance for students with lower reading and writing ability could include sentence starters for each section of the investigation proposal.

Stage 2 engages students in the scientific practices of interacting directly with the natural world (or data drawn from the natural world), using appropriate tools and data collection techniques, and learning that empirical work is not always straightforward—it is an often-ambiguous task that requires adjustments and further data collection as the investigation proceeds. Stage 2 also provides the opportunity for students to see that not all methods are created equal and that some methods work better than others. It is the phenomenon itself and the nature of the question about it that determine how a method is used during a scientific investigation. Stage 2 is complete when students have gathered all the data they need to answer the question.

Stage 3: Data Analysis and Development of a Tentative Argument

The third stage of the instructional model calls for students to develop a tentative argument in response to the guiding question. Each group needs to be encouraged to first "make sense" of the data, measurements, and/or observations collected during stage 2. Once the groups have analyzed and interpreted their data, they can then create their argument. The argument consists of a claim, the evidence they are using to support their claim, and a justification of their evidence (Figure 3). The *claim* is their answer to the guiding question. The *evidence* consists of the data (measurements or observations) they collected, an analysis of the data, and an interpretation of the analysis. In life science, data are commonly analyzed to identify (a) a trend over time, (b) a difference between groups or objects, or (c) a relationship between variables. Students must interpret data in light of the research question, the nature of their study, and the available literature. The justification of the evidence is a statement that defends their choice of evidence by explaining why it is important and relevant, making the concepts or assumptions underlying the analysis and interpretation explicit.

FIGURE 3 _____

Framework for the components of a scientific argument and criteria for evaluating the merits of the argument

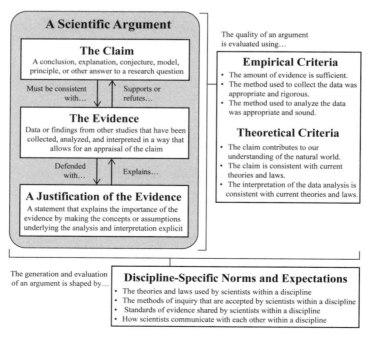

The following example illustrates the three structural components of an argument that was made in response to the guiding question, "Are dolphins more closely related to fish or dogs?"

The claim: Dolphins are more closely related to dogs than to fish.

The evidence: Dolphins and dogs are warm-blooded, get oxygen from the air, and produce milk. Fish are cold-blooded, get the oxygen they need from water, and do not produce milk. Therefore, dolphins and dogs have more in common than dolphins and fish.

A justification of the evidence: Evolution, or descent with modification, indicates that organisms that share a common ancestor will have more traits in common. Organisms that share a common ancestor in the more recent past are more closely related than organisms that share a common ancestor in the distant past.

The claim is the answer given to the guiding question. The individuals making the claim analyzed the data they collected and provided it as the evidence supporting the claim (by highlighting similarities and differences between the groups of organisms) and an interpretation of their analysis (explaining what can be inferred from the observed similarities and differences). Finally, the argument includes a justification of the evidence that makes explicit the underlying concept and assumptions guiding the analysis of the data and the interpretation of the analysis.

It is important for students to understand that, in science, some arguments are better than others. An important aspect of science and scientific argumentation involves the evaluation of the various components of the arguments put forward by others. Therefore, the framework provided in Figure 3 also highlights two types of criteria that students can and should use to evaluate an argument in science: empirical criteria and theoretical criteria. *Empirical criteria* include

- how well the claim fits with all available evidence,
- the sufficiency of the evidence,
- the quality of the evidence,
- the appropriateness of the method used to collect the data, and
- the appropriateness of the method used to analyze the data.

Theoretical criteria refer to standards that are important in science but are not empirical in nature, including

- the sufficiency of the claim (i.e., Does it include everything needed?);

- the usefulness of the claim (i.e., Does it allow us to engage in new inquiries or understand a phenomenon?);

- the consistency of the claim and the reasoning in terms of other accepted theories, laws, or models; and

- the manner in which the data analysis was conducted.

What counts as quality within these different categories, however, varies from discipline to discipline (e.g., biology, physics, geology) and within the specific fields of each discipline (e.g., cell biology, evolutionary biology, genetics). The variation is due to differences in the types of phenomena investigated, what counts as an accepted mode of inquiry (e.g., investigation vs. experimentation), and the theory-laden nature of scientific inquiry. It is important to keep in mind that "what counts" as a quality argument in science is discipline and field dependent.

To allow for the evaluation of the argument, each group of students should create their tentative argument in a medium that can easily be viewed by the other groups. We recommend using a whiteboard or a large piece of butcher paper—at least 2′ × 3′. Students should lay out each component of the argument on the board or paper. Figure 4 shows the general layout for a presentation of an argument and an example of an argument crafted by students. Students can also create their tentative arguments using presentation software such as Microsoft's PowerPoint or Apple's Keynote, with a slide for each component of the

FIGURE 4

The components of an argument that should be included on a whiteboard (outline and example)

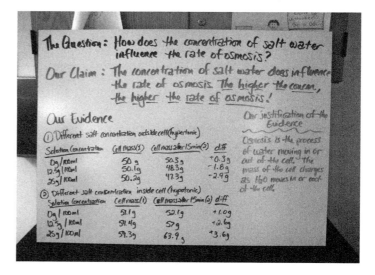

argument. The goal is to select a presentation format that allows students to modify their work as the activity proceeds, so other formats might be acceptable as well.

The intention of stage 3 is to provide the student groups with an opportunity to make sense of what they are seeing or doing during the investigation. As students work together to create a tentative argument, they must talk with each other and determine how to analyze the data and how to best interpret the trends, differences, or relationships that they identify. They must also decide if the evidence (data that have been analyzed and interpreted) that they chose to include in their argument is relevant, sufficient, and convincing enough to support their claim. This, in turn, enables the groups of students to evaluate competing ideas and weed out any claim that is inaccurate, does not fit with all the available data, or contains contradictions.

This stage of the model is challenging for students because they are rarely asked to make sense of a phenomenon based on raw data, so it is important for the teacher to provide active support. This can be a new kind of support role for many teachers, requiring them to question students but not give the answers, even when student frustration may be high. The goal of the teacher is to get students to think about *what* they are doing and *why* they are doing it. Teachers should ask probing questions to remind students about what they are trying to figure out and get them to think about the relevance of their data (e.g., "Why is that characteristic important?") or hold them to careful evaluation of the merits of a tentative idea (e.g., "Does that fit with all the data or what we know about a particular phenomenon?").

Initially students will struggle to develop arguments, often resorting to reliance on inappropriate criteria like plausibility (e.g., "That sounds good to me.") or how the data fit with personal experience (e.g., "But that's what I saw on TV once."). To assist students who may have challenges with reading and evaluating arguments, consider having a whole-class discussion near the beginning of the school year where each element of the argument is discussed in terms of appropriate content. A product of this discussion could be a classroom poster that uses student-generated language to describe the characteristics of each argument element. With practice students will improve and this process will become smoother and take less time. This is an important outcome of the ADI instructional model and helps attain many of the goals of science proficiency.

Stage 4: Argumentation Session

In this stage, each group is given an opportunity to share, evaluate, and revise their tentative arguments with the other groups. This stage is probably least familiar to many teachers, but it is critical because scientific argumentation (i.e., arguing from evidence) is an essential practice in science that drives scientific fields forward. Critique of our explanations and arguments leads to better outcomes, and students learn more about the content and develop better critical thinking skills when they are exposed to the alternative ideas, respond to the questions and challenges of other students, and evaluate the merits of competing ideas (Duschl, Schweingruber, and Shouse 2007; NRC 2012). This stage also enables students to learn how to distinguish between ideas by using rigorous scientific criteria and to develop scientific habits of mind (e.g., treating ideas with initial skepticism, insisting that the reasoning and assumptions be made explicit, and insisting that claims be supported by valid evidence).

It is critical to keep in mind that this type of classroom discussion is foreign to most students and requires special support from the teacher. Having students generate their arguments in a public medium (such as the whiteboard) where others can see them is one such support; it helps students focus their attention on evaluating evidence rather than attacking the source of the ideas.

To allow all of the groups to share their arguments, we recommend that one member of each group stay at that group's lab station to share the group's argument while the other members of that group go to the other lab stations one at a time to listen to and critique the arguments developed by their classmates (see Figure 5). Students should *not* know ahead of time which group member will be called upon to stay at the whiteboard to share the group's ideas; this helps ensure better participation by all group members in the data analysis and development of the argument. This type of format ensures that all ideas are heard, more students are actively involved in the process, and ideas become more refined.

Just as is the case in earlier stages, it is important for the teacher to be involved in (without leading) the discussions during the argumentation session. Once again, the teacher should move from group to group to keep students on task and model good scientific argumentation. The teacher can ask the presenter questions such as, "How did you analyze the available data?" or "Were there any data that did not fit with your claim?" to encourage students to use empirical criteria to evaluate the quality of the arguments. The teacher can also ask the presenters to explain how the claims they are presenting fit with the theories, laws, or models of science or to explain why the evidence they used is important. In addition, the teacher can also ask the students who are listening to the presentation questions such as "Do you think their analysis is accurate?" or "Do you think their interpretation is sound?" or even "Do you think their claim fits with what we know about X?" The purpose of the teacher's questions is to remind students to use empirical and theoretical criteria to evaluate an argument during the discussions. Overall, the goal of the teacher at this stage of the lesson is to encourage students to think about how they know what they know and why some claims are more valid or acceptable in science. The teacher is shifting the job of sense-making to the students, so it is important to refrain from telling students that they are right or wrong at this stage of the model.

Stage 5: Explicit and Reflective Discussion

This stage of the ADI model provides a context for teachers to explain the nature of scientific knowledge and how this knowledge develops over time. Current research suggests that students only develop an appropriate understanding of the nature of science (NOS) and the nature of scientific inquiry (NOSI) when teachers discuss these concepts in an explicit fashion (Abd-El-Khalick and Lederman 2000; Akerson, Abd-El-Khalick, and Lederman 2000; Lederman and Lederman 2004; Lederman et al. 2014; Schwartz, Lederman, and Crawford 2004). In stage 5, the original student groups reconvene and discuss what they learned by interacting with individuals from the other groups during the argumentation session. Afterward, students can modify their tentative argument as needed or conduct an additional analysis of the data.

FIGURE 5

Students participate in the argumentation session

After their modifications are complete, the teacher should lead a whole-class discussion in which several students from different groups are encouraged to explain what they learned about the phenomenon under investigation. The teacher's role is to guide the class toward a scientifically acceptable conclusion through careful questioning. It is important to remember that this is a *discussion*, not a lecture. The teacher can also discuss any issues that were a common challenge for the groups during stages 2 and 3 of the activity.

Teachers should discuss one or two crosscutting concepts (e.g., the relationship between structure and function, or the importance of identifying and explaining patterns in nature) during this stage; the students' experiences during the lab investigation can be used as a concrete example of each concept. In addition, teachers should discuss one or two NOS (e.g., the difference between laws and theories, or the role that creativity and imagination play in science) or NOSI (e.g., what does and does not count as an experiment in science, or the different methods that scientists use to answer different types of questions) concepts that were prominent in the activity, again using the students' experiences during the investigation to illustrate these important concepts.

Stage 6: Writing the Investigation Report

Stage 6 is included in the ADI model because writing is an important part of doing science. Scientists must be able to read and understand the writing of others as well as evaluate its worth. They also must be able to share the results of their own research through writing. In addition, writing helps students learn how to articulate their thinking in a clear and concise manner, encourages metacognition, and improves student understanding of the content (Wallace, Hand, and Prain 2004). Finally, and perhaps most important, writing makes each student's thinking visible to the teacher (which facilitates assessment) and enables the teacher to provide students with the educative feedback they need in order to improve.

In stage 6, each student is required to write an individual investigation report using his or her group's argument. The report, which can be written during class or can be assigned as homework, should be centered on three fundamental questions:

1. What were you trying to do and why?

2. What did you do and why?

3. What is your argument?

An important component of this process is to encourage students to use tables or graphs to help organize the data they gathered and require them to reference the tables or graphs in the body of the report. This method allows them to learn how to construct an explanation, argue from evidence, and communicate information. It also enables students to master the disciplinary-based writing skills outlined in the *CCSS ELA* (NGAC and CCSSO 2010). The report can be written during class or assigned as homework.

The format of the report is designed to emphasize the persuasive nature of science writing and to help students learn how to communicate in multiple modes (words, figures, tables, and equations or formulas). The three-question format is well aligned with the components of a traditional lab report (i.e., introduction, procedure, results and discussion) but allows students to see the important role argument plays in science. We strongly recommend that you *limit the length of the investigation report* to two double-spaced pages or one single-spaced page. This limitation encourages students to write in a clear and concise manner, because there is little room for extraneous information. Requiring a short report is less intimidating to students than requiring a lengthier report, and it lessens the work required in the subsequent stages.

Many students face challenges in writing these reports, not only attributable to their understanding of the content they are writing about but also to their skill in crafting written products that reflect their understanding. To help these students, teachers should consider some different ways to scaffold the writing process when the students are doing their first or second ADI activity. A variety of approaches have been used, beginning with simply allowing students to write their reports in class and being present to provide individual assistance. Another approach involves stopping after stage 1 and having students write the first section of the report. Then, during stage 3, when students are analyzing their data, the teacher takes class time to have students write the second section of the report. Finally, after stage 4, the only section left for students to write is the final argument section of the report. Scaffolding strategies may be used in the beginning of the school year, but you should consistently remove some support each time; otherwise students will come to expect that support routinely and limit their learning of writing skills. Also, you should seek out the assistance of colleagues, including reading/writing coaches, to see what strategies they have found successful with the students at that school.

Stage 7: Double-Blind Group Peer Review

In this stage, each student is required to submit to the teacher two to four typed copies of the investigation report. Students do not place their names on the reports; instead, each student uses an identification number (assigned by the teacher) to maintain anonymity; this ensures that reviews are based on the ideas presented and not the person presenting the ideas. The students once again work in their original groups, and each group receives three or four sets of reports (i.e., the reports written by three or four different authors from other groups) and one peer-review guide for each author's report (see Appendix 4, p. 365). The peer-review guide lists specific criteria that are to be used by the group as they cooperatively evaluate the quality of each section of the investigation report as well as the mechanics of the writing. There is also space for the reviewers to provide the author with feedback about how to improve the report.

Reviewing each report as a group is an important component of the peer-review process because it provides students with a forum to discuss "what counts" as high quality or acceptable and in so doing forces them to reach a consensus during the process. This method also helps prevent students from checking off "yes" for each criterion on the peer-review guide without thorough consideration of the merits of the paper. It is also important for students to provide constructive and specific feedback to the author when areas of the paper are found to not meet the standards established by the peer-review guide. The peer-review process provides students with an opportunity to read good and bad examples of reports. This helps the students learn new ways to organize and present information, which in turn will help them write better on subsequent reports. (*Note:* You will have access to the peer-review sheets when scoring the papers, and you should hold groups accountable if they are not doing their peer-review job properly.) To encourage students' improvement through this process, you should limit the amount of direct feedback that you provide.

Writing is not the only literacy skill that is supported during stage 7—students also develop science reading skills. For example, students must be able to determine the central ideas or conclusions of a text and determine the meaning of symbols, key terms, and other domain-specific words. In addition, they must be able to assess the reasoning and evidence that an author includes in a text to support his or her claim and to compare or contrast findings presented in a text to those from other sources. Students can develop all these skills, as well as the other discipline-based reading standards found in the *CCSS ELA*, when they are required to read and critically review reports written by their classmates.

Teachers can adapt the peer-review guide to a specific ADI activity by changing the language of the rubric items to include targeted items that address ideas and phenomena experienced during that activity. Also, explicit instruction about what constructive critique of writing looks like will be helpful for all students in their development of this skill.

Stage 8: Revision and Submission of the Investigation Report

The final stage in the ADI instructional model is to revise the report based on the suggestions given during the peer review. If the report met all the criteria, the student may submit it to the teacher without revision. Students whose reports are found by the peer-review group to be acceptable also have the option to revise the reports if they so desire after reviewing the work of other students. However, if a report was found unacceptable by the group during peer review, the author is required to rewrite it using the reviewers' comments and suggestions as a guideline. Once the report is revised, it is turned in to the teacher for evaluation. The author is required to explain what he or she did to improve each section of the report in response to the reviewers' suggestions (or explain why the author decided to ignore the reviewers' suggestion); this is done on the peer-review sheet in the space provided for author response. *All* submissions must include the "rough draft" and the peer-review sheet. The teacher then scores the report in the instructor score column of the peer-review sheet, and these ratings are used to assign an overall grade.

The ADI model has many positive outcomes. Students improve their writing skills and reading skills, and they further develop their reasoning skills. Their understanding of the instructional content is deepened. Finally, students have the opportunity to improve their academic performance (i.e., final paper score).

The Role of the Teacher During Argument-Driven Inquiry

If the ADI instructional model is to be successful and student learning is to be optimized, the role of the teacher during a lab activity designed using this model must be different than the teacher's role during a more traditional lab. The teacher must act as a resource for the students—rather than as a director—as they work through each stage of the activity, encouraging students to think about *what they are doing* and *why they made that decision* throughout the process. Teachers must avoid "telling" in response to student questions— and students *will* press for information. Instead, teachers must support student thinking and sense-making through probing questions. For example, teachers might ask, "Why do you want to set up your equipment that way?" or "What type of data will you need to collect to be able to answer that question?" Teachers must restrain from telling or showing students how to "properly" conduct the investigation. Quality control is key: Teachers have to require high standards from students for their methodology, helping them to learn what counts as good scientific technique and what counts as a strong scientific argument.

Students will struggle and will be frustrated, and this is especially true when ADI is new to them. Teachers will have to accept that a certain amount of this is good for learning and that our tendency to want to help alleviate problems is not as helpful for deep learning. Students must be allowed to try things that will fail—this is a tremendous learning opportunity. The teacher's role is to help them learn from failure both for the specific situation and for future, more general situations. The tendency to make ADI activities "student-proof" by providing additional directions does not serve the best interests of the students. Students learn as much, and probably more, from failing, making adjustments, and finally succeeding than they do if everything goes right the first time. This kind of learning is strengthened by the fact that in ADI lab activities in which students design a poor method to collect data or analyze their results in an inappropriate manner, it is their *classmates* who identify the mistakes during the argumentation session. This leads to increased learning by both groups of students, and teachers will encounter many teachable moments in these instances. These learning experiences can be quite powerful for all.

Because the teacher's role in an ADI lab is different from what typically happens in laboratories, we have created a chart describing teacher behaviors that are consistent and inconsistent with each stage of the instructional model (see Table 2, pp. 16–17). This table is organized by stage because what the students and the teacher need to accomplish during each stage is different. It might be helpful to keep this table handy as a guide when first attempting to implement the lab activities found in the book. When using the table, consider how the teacher's role in ADI differs from that in more traditional lab instruction.

TABLE 2

Teacher behaviors during the stages of the ADI instructional model

Stage	What the teacher does that is ...	
	Consistent with ADI model	**Inconsistent with ADI model**
1: Identification of the task and the guiding question; "tool talk"	• Sparks students' curiosity • "Creates a need" for students to design and carry out an investigation • Organizes students into collaborative groups • Supplies students with the materials they will need • Holds a "tool talk" to show students how to use equipment or to illustrate proper technique • Provides students with hints	• Provides students with possible answers to the research question • Tells students that there is one correct answer • Tells students what they "should expect to see" or what results "they should get"
2: Designing a method and collecting data	• Encourages students to ask questions as they design their investigations • Asks groups questions about their method (e.g., "Why did you do it this way?") and the type of data they expect from that design • Reminds students of the importance of specificity when completing their investigation proposals	• Gives students a procedure to follow • Does not question students about their method or the type of data they expect to collect • Approves vague or incomplete investigation proposals
3: Data analysis and development of a tentative argument	• Reminds students of the research question and what counts as appropriate evidence in science • Requires students to generate an argument that provides and supports a claim with genuine evidence • Asks students what opposing ideas or rebuttals they might anticipate • Provides related theories and reference materials as tools	• Requires only one student to be prepared to discuss the argument • Moves to groups to check on progress without asking students questions about why they are doing what they are doing • Does not interact with students (uses the time to catch up on other responsibilities) • Tells students the right answer
4: Argumentation session	• Reminds students of appropriate behaviors in the learning community • Encourages students to ask questions of peers • Keeps the discussion focused on the elements of the argument • Encourages students to use appropriate criteria for determining what does and does not count	• Allows students to negatively respond to others • Asks questions about students' claims before other students can ask • Allows students to discuss ideas that are not supported by evidence • Allows students to use inappropriate criteria for determining what does and does not count

Table 2 *(continued)*

Stage	What the teacher does that is ...	
	Consistent with ADI model	**Inconsistent with ADI model**
5: Explicit and reflective discussion	• Encourages students to discuss what they learned about the content and how they know what they know • Encourages students to discuss what they learned about the nature of science • Encourages students to think of ways to be more productive next time	• Provides a lecture on the content • Skips over the discussion about the nature of science and the nature of scientific inquiry to save time • Tells students "what they should have learned" or "this is what you all should have figured out"
6: Writing the investigation report	• Reminds students about the audience, topic, and purpose of the report • Provides the peer-review guide in advance • Provides an example of a good report and an example of a bad report	• Has students write only a portion of the report • Moves on to the next activity/topic without providing feedback
7: Double-blind group peer review	• Reminds students of appropriate behaviors for the review process • Ensures that all groups are giving a quality and fair peer review to the best of their ability • Encourages students to remember that while grammar and punctuation are important, the main goal is an acceptable scientific claim with supporting evidence and justification • Holds the reviewers accountable	• Allows students to make critical comments about the author (e.g., "This person is stupid") rather than their work (e.g., "This claim needs to be supported by evidence") • Allows students to just check off "Yes" on each item without providing a critical evaluation of the report
8: Revision and submission of the investigation report	• Requires students to edit their reports based on the reviewers' comments • Requires students to respond to the reviewers' ratings and comments • Has students complete the checkout questions after they have turned in their report	• Allows students to turn in a report without a completed peer-review guide • Allows students to turn in a report without revising it first

References

Abd-El-Khalick, F., and N. G. Lederman. 2000. Improving science teachers' conceptions of nature of science: A critical review of the literature. *International Journal of Science Education* 22: 665–701.

Akerson, V., F. Abd-El-Khalick, and N. Lederman. 2000. Influence of a reflective explicit activity-based approach on elementary teachers' conception of nature of science. *Journal of Research in Science Teaching* 37 (4): 295–317.

Duschl, R. A., H. A. Schweingruber, and A. W. Shouse, eds. 2007. *Taking science to school: Learning and teaching science in grades K-8*. Washington, DC: National Academies Press.

Lederman, N. G., and J. S. Lederman. 2004. Revising instruction to teach the nature of science. *The Science Teacher* 71 (9): 36–39.

Lederman, J., N. Lederman, S. Bartos, S. Bartels, A. Meyer, and R. Schwartz. 2014. Meaningful assessment of learners' understanding about scientific inquiry: The Views About Scientific Inquiry (VASI) questionnaire. *Journal of Research in Science Teaching* 51 (1): 65–83.

National Governors Association Center for Best Practices and Council of Chief State School Officers (NGAC and CCSSO). 2010. *Common core state standards*. Washington, DC: NGAC and CCSSO.

National Research Council (NRC). 2012. *A framework for K–12 science education: Practices, crosscutting concepts, and core ideas*. Washington, DC: National Academies Press.

Schwartz, R. S., N. Lederman, and B. Crawford. 2004. Developing views of nature of science in an authentic context: An explicit approach to bridging the gap between nature of science and scientific inquiry. *Science Education* 88: 610–645.

Wallace, C., B. Hand, and V. Prain, eds. 2004. *Writing and learning in the science classroom*. Boston: Kluwer Academic Publishers.

CHAPTER 2
Lab Investigations

This book includes 20 life science lab investigations designed around the argument-driven inquiry (ADI) instructional model. The investigations are not meant to constitute a curriculum or to replace an existing one but rather to change the laboratory component of middle school life science courses. A teacher can use these investigations as a way to introduce students to new content ("introduction labs") or as a way to give students an opportunity to apply a theory, law, or unifying concept introduced in class in a novel situation ("application labs"). To facilitate curriculum and lesson planning, the lab investigations have been aligned with *A Framework for K–12 Science Education* (National Research Council [NRC] 2012); the *Common Core State Standards* (NGAC and CCSSO) in English language arts (*CCSS ELA*) and mathematics (*CCSS Mathematics*); and various aspects of the nature of science (NOS) and the nature of scientific inquiry (NOSI; Abd-El-Khalick and Lederman 2000; Lederman et al. 2002; Lederman et al. 2014). The matrixes in Appendix 1 (p. 349) summarize these alignments.

Many of the ideas for the investigations in this book came from existing resources; however, we modified the activities from those resources to fit with the focus and nature of the ADI instructional model. This model provides teachers with a way to transform classic or traditional lab activities into authentic and educative investigations that enable students to become more proficient in science.

Each ADI activity in this book has been tested several times in middle school classrooms. Activity development followed a regular sequence. The ideas for the activity often came from existing resources that were then modified to fit the ADI instructional model. Once the lab was written, several practicing biologists reviewed the activity for content accuracy. Next, all activities were tested in all types of sections of middle school life science courses (including general and honors sections). Teacher feedback was collected on each activity and student learning gains reviewed from each situation. This feedback was used to modify each activity to improve its efficacy in the classroom. Following this first cycle of testing and modification, a second cycle of testing was completed (i.e., pilot testing in middle school classes, collection of feedback and learning gains, and activity modification). The results of this second round of testing constitute the lab investigations found in this book.

These lab investigations were developed as part of a three-year research project funded by the Institute of Education Sciences through grant R305A100909. The goals of this project were to develop a set of ADI lab activities with teachers and then have the teachers use the activities in their classrooms to refine the ADI instructional model and learn more about its effects on students. The project examined what students learn when they complete eight or more ADI labs over the course of a school year. The initial testing site for the project was Florida State University Schools (a K–12 laboratory school), and the activities were subsequently tested in other public school classrooms. This research indicates that not only do students make substantial gains in their understanding of important content and NOS after participating in at least eight ADI investigations, but they also have much better inquiry and writing skills (Grooms, Sampson, and Carafano 2012; Sampson, Grooms, and Enderle 2012; Sampson et al. 2013). To learn more about the research associated with the ADI instructional model, visit *www.argumentdriveninquiry.com.*

Teacher Notes

Each teacher must decide when and how to use a lab to best support student learning. To help with this decision making, we have included Teacher Notes for each investigation. These notes include information about the purpose of the lab, the time needed to implement each stage of the model for that lab, the materials needed, and hints for implementation. There is also a "Topic Connections" section that shows how each ADI lab investigation is aligned with the NRC *Framework*, the NOS or NOSI concepts, and the *CCSS ELA* and/or the *CCSS Mathematics.* In the following subsections, we will describe the information provided in each section of the Teacher Notes.

Purpose

This section of the Teacher Notes describes the main idea of the lab and indicates whether the activity is an introduction lab or an application lab. In either case, very little emphasis needs to be placed on the vocabulary or on making sure that students "know their stuff" before the lab investigation begins. Instead, with the combination of the information provided in the Lab Handout, students' evolving understanding of the actual practice of science, and the various resources available to the students (e.g., the science textbook, the internet, and, of course, the teacher), students will develop a better understanding of the content as they work through the activity. This section also highlights the NOS or NOSI concepts that should be discussed during the explicit and reflective discussion stage.

The Content

This section of the Teacher Notes provides an overview of the concept that is being introduced to the students or that students will need to apply during the investigation. It also provides an answer to the guiding question of the investigation.

Timeline

Unlike most traditional laboratories, which can be completed in a single class period, ADI labs typically take three to five instructional days complete. More time may be needed for the first few labs that your students conduct, but the time needed will be reduced as they become familiar with the ADI process. The "Timeline" section describes the instructional time (presented as a range) needed to complete each lab and refers to timeline options more fully described in Appendix 2 (p. 355). The figures in Appendix 2 show the day and stage(s) of the ADI model that ideally should be completed in class each day and outline the resulting products of each stage.

It is important to note that although the days are listed chronologically in the timeline options in Appendix 2, they do not necessarily have to fall on consecutive days. For example, when students need to write an investigation report, the teacher can allow them to have more than one night to complete the work, especially when they are getting used to what is expected of them. Also, some of the lab stages do not take an entire class period to complete, especially once students are acclimated to the ADI model. As with all teaching, flexibility is key!

Materials and Preparation

This section of the Teacher Notes describes the lab supplies and instructional materials (e.g., Lab Handout, investigation proposal, and peer-review guide) needed to implement the lab activity. To help teachers plan for different-sized classes, this section lists quantities needed for each group or each student. When resources are scarce, groups can share materials. We have also included specific suggestions for some lab supplies that were found to work best during the pilot tests. However, substitutions can be made if needed. Be sure to test all lab supplies before conducting the lab with the students, because new materials often have unexpected consequences.

We also explain in this section how to prepare the materials (including consumables) that the students will use during the investigation. Some labs require an hour or more of preparation time, so teachers should review this section at least 24 hours before conducting the investigation in the classroom.

Safety Precautions

This section of the Teacher Notes provides an overview of potential safety hazards as well as safety protocols that should be followed to make the laboratory safer for students. These are based on legal safety standards and current safety practices. Teachers should review and follow all local policies and protocols used within their school district and/or school (e.g., the district chemical hygiene plan, Board of Education safety policies).

Topics for the Explicit and Reflective Discussion

This section provides an overview of the content at the heart of the investigation, relevant crosscutting concepts, and NOS/NOSI concepts. It also provides advice for teachers on how to encourage students to reflect on the strengths and limitations of their investigations and how to improve the design of their investigations in the future.

Hints for Implementing the Lab

These lab investigations have been tested by many teachers many times, and we have collected "hints" from these teachers for each stage of the ADI process. These hints can help you avoid some of the pitfalls teachers experienced during the testing and should make the investigation run smoothly. Tips for making the investigation safer are also included, where appropriate.

Topic Connections

This section, which is designed to inform curriculum and lesson planning, includes a table that highlights the scientific practices, crosscutting concepts, and core ideas from the NRC *Framework* that are aligned with the lab activity. The table also outlines supporting ideas, the NOS/NOSI concepts, and the *CCSS ELA* and the *CCSS Mathematics* addressed by the activity.

Instructional Materials

The instructional materials included in this book are reproducible copy masters that are designed to support students as they participate in an ADI lab investigation. The materials include Lab Handouts, Lab Reference Sheets, investigation proposals, the peer-review guide, and Checkout Questions. In the following subsections, we will provide an overview of these important materials.

Lab Handout

At the beginning of each lab investigation, each student should be given a copy of the Lab Handout. This handout provides information about the phenomenon that the students will investigate and a guiding question for the students to answer. The handout also provides hints for student to help them design their investigation in the Getting Started section, information about what to include in their tentative argument, safety precautions, and the requirements for the investigation report.

In light of page limits for school copiers, there are condensed versions of the student handouts that are only a few pages in length; these condensed versions are available at *www.nsta.org/publications/press/extras/adi-lifesciences.aspx*.

Lab Reference Sheet

Some lab investigations include an optional Lab Reference Sheet that provides additional information that the students can use as part of their investigation. If a teacher decides to use the Lab Reference Sheet, we recommend making a class set or providing one copy to each lab group.

Investigation Proposal

To help students design better investigations, we have developed and included three different types of investigation proposals in this book (see Appendix 3, p. 361). These investigation proposals are optional, but we have found that students design and carry out much better investigations when they are required to fill out a proposal and then get teacher feedback about their method before they begin. We provide recommendations in the Teacher Notes about which investigation proposal (A, B, or C) to use for a particular lab. If a teacher decides to use an investigation proposal as part of a lab, we recommend providing one copy for each group.

The Lab Handout for students has a heading asking if an investigation proposal is required, followed by boxes to check "yes" or "no." Teachers should make sure that students check the appropriate box when introducing the lab activity.

Peer-Review Guide

The peer-review guide (see Appendix 4, p. 365) is designed to make explicit the criteria used to judge the quality of an investigation report. We recommend that teachers make one copy for each student and then provide the copies to the students before they begin writing their investigation reports. This will ensure that students understand how they will be evaluated. During stage 7 of the model (double-blind group peer review), each group should fill out the peer-review guide as they review the reports of their classmates (each group will need to review three or four different reports). The reviewers should rate the report on each criterion and provide advice to the author about ways to improve. Once the review is complete, the author needs to revise the report and respond to the reviewers' ratings and comments in the appropriate sections of the peer-review guide. The completed peer-review guide should be submitted to the teacher along with the final and first draft of the report for a final evaluation. To score the report, the teacher can simply fill out the "Instructor Score" column of the rubric and then total the scores.

Checkout Questions

To facilitate classroom assessment, we have included a set of Checkout Questions for each lab investigation. The questions target the life sciences core ideas, the crosscutting concepts, and the NOS/NOSI concepts addressed in the lab. Students should answer the questions on the same day they turn in their final reports. One copy of the Checkout Questions is needed for each student. The students should complete these questions on their own. The teacher can use the students' responses, along with the report, to determine if the students learned what they needed to during the lab, and then reteach as needed.

References

Abd-El-Khalick, F., and N. G. Lederman. 2000. Improving science teachers' conceptions of nature of science: A critical review of the literature. *International Journal of Science Education* 22: 665–701.

Grooms, J., V. Sampson, and P. Carafano. 2012. The impact of a new instructional model on high school science writing. Paper presented at the annual international conference of the American Educational Research Association.

Lederman, N. G., F. Abd-El-Khalick, R. L. Bell, and R. S. Schwartz. 2002. Views of nature of science questionnaire: Toward a valid and meaningful assessment of learners' conceptions of nature of science. *Journal of Research in Science Teaching* 39 (6): 497–521.

Lederman, J., N. Lederman, S. Bartos, S. Bartels, A. Meyer, and R. Schwartz. 2014. Meaningful assessment of learners' understanding about scientific inquiry: The Views About Scientific Inquiry (VASI) questionnaire. *Journal of Research in Science Teaching* 51 (1): 65–83.

National Governors Association Center for Best Practices and Council of Chief State School Officers (NGAC and CCSSO). 2010. *Common core state standards*. Washington, DC: NGAC and CCSSO.

National Research Council (NRC). 2012. *A framework for K–12 science education: Practices, crosscutting concepts, and core ideas*. Washington, DC: National Academies Press.

Sampson, V., P. Enderle, J. Grooms, and S. Witte. 2013. Writing to learn and learning to write during the school science laboratory: Helping middle and high school students develop argumentative writing skills as they learn core ideas. *Science Education* 97 (5): 643–670.

Sampson, V., J. Grooms, and P. Enderle. 2012. Using laboratory activities that emphasize argumentation and argument to help high school students learn how to engage in scientific inquiry and understand the nature of scientific inquiry. Paper presented at the annual international conference of the National Association for Research in Science Teaching.

SECTION 2

Life Sciences Core Idea 1

From Molecules to Organisms: Structures and Processes

Introduction Labs

LAB 1

Teacher Notes

Lab 1. Cellular Respiration: Do Plants Use Cellular Respiration to Produce Energy?

Purpose

The purpose of this lab is to *introduce* students to biochemical processes at the cellular level. Specifically, this investigation gives students an opportunity to explore the process of cellular respiration that provides energy for cellular activities. Teachers will need to help students understand that oxygen gas (O_2) is used by cellular respiration and carbon dioxide gas (CO_2) is given off by it, because the concentration of these gases is how students will measure the cellular process. This lab gives students an opportunity to learn about cause-and-effect relationships and to understand how energy and matter flow through living systems. Students will also have the opportunity to reflect on the difference between observations and inferences and the difference between data and evidence in science.

The Content

One characteristic of living things is they must take in nutrients and give off waste in order to survive. This is because all living tissues (which are made of cells) are constantly using energy. In animals, this energy comes from a reaction called *cellular respiration,* which refers to a process that occurs inside cells where sugar is used as a fuel source. This process happens in a specific location inside cells called the *mitochondrion* (the plural form is *mitochondria*), which is an organelle inside of cells that is mainly responsible for producing energy for the activities of the cell. Figure 1.1 provides both a drawing and an electron microscopic image of a mitochondrion. Mitochondria are found in both plant and animal cells.

FIGURE 1.1

(a) Drawing and (b) electron microscopic image of a mitochondrion

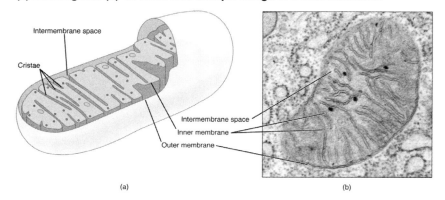

Intermembrane space

Cristae

Intermembrane space

Inner membrane

Outer membrane

(a)

(b)

Scientists have determined that mitochondria used to be separate organisms rather than part of cells. A major piece of evidence supporting this conclusion is that these organelles contain their own DNA, which codes for several proteins that are involved in the process of cellular respiration. Scientists think that early bacteria living billions of years ago absorbed mitochondrial ancestors into their cells. As mitochondria became better at providing energy for the bacteria to use, they provided a survival advantage to the bacteria that absorbed them. In return, the bacteria provided a safe habitat for the mitochondria to live and function. Over millions of years, the mitochondria became fixed organelles in these bacteria, with much of their DNA transferring to the nuclei of the bacteria where they were located.

The energy that cells use comes from the chemical bonds of sugar. During cellular respiration, oxygen helps convert the chemical energy in sugar molecules into a form animals can use. The energy from these molecules gets transferred by moving electrons from one molecule to another. When electrons are added or taken away, new chemical bonds and types of molecules can be formed. The oxygen helps transfer electrons from the chemical bonds in sugar to the chemical bonds in another molecule. Most living organisms use a special molecule known as *ATP* (adenosine triphosphate) to provide energy for all the activities taking place in their cells.

The following equation describes the chemical changes that occur during cellular respiration:

Sugar ($C_6H_{12}O_6$) + oxygen (O_2) → water (H_2O) + carbon dioxide (CO_2) + usable energy (ATP)

There are three stages of cellular respiration, which involves three distinct sets of reactions: glycolysis, the citric acid cycle, and the electron transport system (Figure 1.2). The first stage is known as *glycolysis,* which takes place outside of the mitochondria. Glycolysis involves breaking a sugar molecule into smaller pieces that can more easily cross the mitochondrial membrane. A molecule of ATP is made during this stage as well. ATP is made by recycling ADP (adenosine diphosphate), which is present in the cell, and adding a third phosphate molecule (inorganic phosphate). A lot of energy is stored in the bond between the second and third phosphates. To release the energy in that bond, different cellular reactions and processes break the third phosphate molecule off of the ATP molecule. Glycolysis is a very simple way in which cells, including certain bacteria, can make energy without O_2 or mitochondria.

The second stage of cellular respiration is the *citric acid cycle.* This cycle of reactions takes reshaped carbon chain molecules, which were originally sugar, and changes their form in several different ways. During these reactions, further energy is released through breaking bonds in the carbon chains and using that energy to create a bond between an ADP and a phosphate molecule.

FIGURE 1.2 _____

Stages of cellular respiration and the locations of the stages in relation to the mitochondria

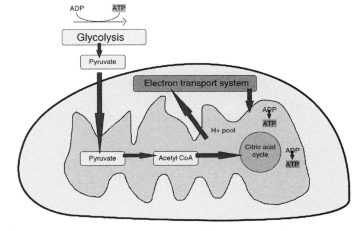

Note: Acetyl CoA = acetyl coenzyme A; ADP = adenosine diphosphate; ATP = adenosine triphosphate.

These reactions also release a lot of H⁺ ions inside the mitochondria that are important for the next set of reactions. As a cycle, the citric acid reactions start and end with the same carbon chain molecule, so they can continue to run through the same set of reactions. As the carbon chains get rearranged during the cycle, a molecule of CO_2 is given off as a waste product.

The third set of reactions for cellular respiration is the *electron transport system* (also called the *electron transport chain*). This "chain" refers to a series of proteins embedded in the inside membrane of the mitochondria. These proteins take energized electrons from the citric acid cycle and transfer them down the protein chain. These reactions are where oxygen becomes necessary. Oxygen has one of the strongest "pulls" of most elements on electrons. This attraction by oxygen helps pull the electrons down the electron transport chain. At the end of the chain, the oxygen adds the electron along with some H⁺ ions to give off a molecule of H_2O. The energized electron loses a small amount of energy during each transfer in the chain. That lost energy pushes H⁺ ions across the inside membrane into the intermembrane space, building up a high concentration there. Molecules naturally move from an area of high concentration to an area of low concentration. The only way for an H⁺ ion to move back across the inside membrane is through a special enzyme channel known as ATP synthase, which is an enzyme that can easily bond ADP with a phosphate group. As each H⁺ ion moves through ATP synthase, the enzyme is activated and creates another molecule of ATP.

Because cellular respiration uses O_2 and produces CO_2, it is as if a cell is "breathing" during the process. Indeed, cellular respiration is connected to a human body's cardiovascular respiration. Cardiovascular respiration brings in the O_2 needed for cellular respiration and dissolves it into blood that carries the O_2 to cells. The CO_2 released by cellular respiration is absorbed back into the blood, where it is carried back to the lungs and exhaled out of the body. We know that humans use respiration to produce energy because when a human breathes, the air that he or she inhales contains about 21% O_2 and less than 1% CO_2; however, when he or she exhales, the air contains about 15% O_2 and 5% CO_2.

Plants also use cellular respiration. Because plants are the main organisms that go through the process of photosynthesis, many people have a misconception that plants use photosynthesis for their energy needs. It is easy to confuse these processes because the products of one become the reactants of the other. However, mitochondria are present in both plant and animal cells, which means that cellular respiration occurs in both types of cells because that is the major activity of those organelles.

Timeline

The instructional time needed to implement this lab investigation is 180–250 minutes. Appendix 2 (p. 355) provides options for implementing this lab investigation over several class periods. Option A (250 minutes) should be used if students are unfamiliar with scientific writing, because this option provides extra instructional time for scaffolding the writing process. You can scaffold the writing process by modeling, providing examples, and providing hints as students write each section of the report. Option B (180 minutes) should be used if students are familiar with scientific writing and have the skills needed to write an investigation report on their own. In option B, students complete stage 6 (writing the investigation report) and stage 8 (revising the investigation report) as homework.

Materials and Preparation

The materials needed to implement this investigation are listed in Table 1.1. You will need laptop computers to connect to and collect data from the sensors. These sensors can be ordered from a science supply company such as Vernier or Pasco. If you do not have both sensors available, then use the one you have and ask students to consider if a decrease or increase in the gas indicates respiration activity.

TABLE 1.1

Materials list

Item	Quantity
Consumables	
Germinating peas (soaked in water and beginning to sprout)	30 per group
Dry peas	30 per group
Plastic beads	30 per group
Equipment and other materials	
O_2 gas sensor	1 per group
CO_2 gas sensor	1 per group
Biochambers (or sealed containers that can accommodate gas sensors)	3 per group
Go!Link adaptor and laptop computer	1 per group
Sanitized indirectly vented chemical-splash goggles	1 per student
Chemical-resistant apron	1 per student
Gloves	1 pair per student
Investigation Proposal A (optional)	1 per group
Whiteboard, 2' × 3'*	1 per group
Lab Handout	1 per student
Peer-review guide	1 per student
Checkout Questions	1 per student

* As an alternative, students can use computer and presentation software such as Microsoft PowerPoint or Apple Keynote to create their arguments.

LAB 1

Safety Precautions

Follow all normal lab safety rules. In addition, take the following safety precautions:

1. Put on sanitized indirectly vented chemical-splash goggles and laboratory apron and gloves before starting the lab activity.

2. Wash hands with soap and water after completing the lab activity.

Topics for the Explicit and Reflective Discussion

Concepts That Can Be Used to Justify the Evidence

To provide an adequate justification of their evidence, students must explain why they included the evidence in their arguments and make the assumptions underlying their analysis and interpretation of the data explicit. In this investigation, students can use the following concepts to help justify their evidence:

- Cellular respiration is the chemical process that produces energy for cells.
- Cellular respiration takes place in mitochondria.
- Mitochondria are present in animal and plant cells.
- Cellular respiration uses oxygen and gives off carbon dioxide.

We recommend that you review these concepts during the explicit and reflective discussion to help students make this connection.

How to Design Better Investigations

It is important for students to reflect on the strengths and weaknesses of the investigation they designed during the explicit and reflective discussion. Students should therefore be encouraged to discuss ways to eliminate potential flaws, measurement errors, or sources of bias in their investigations. To help students be more reflective about the design of their investigation, you can ask the following questions:

- What were some of the strengths of your investigation? What made it scientific?
- What were some of the weaknesses of your investigation? What made it less scientific?
- If you were to do this investigation again, what would you do to address the weaknesses in your investigation? What could you do to make it more scientific?

Crosscutting Concepts

This investigation is aligned with two crosscutting concepts found in *A Framework for K–12 Science Education,* and you should review these concepts during the explicit and reflective discussion.

- *Cause and effect: Mechanism and explanation:* Natural phenomena have causes, and uncovering causal relationships (e.g., how changes in environmental conditions influence the size of cells) is a major activity of science.

- *Energy and matter: Flows, cycles, and conservation:* In science it is important to track how energy and matter move into, out of, and within systems.

The Nature of Science and the Nature of Scientific Inquiry

This investigation is aligned with two important concepts related to the *nature of science* (NOS) and the *nature of scientific inquiry* (NOSI), and you should review these concepts during the explicit and reflective discussion.

- *The difference between observations and inferences:* An observation is a descriptive statement about a natural phenomenon, whereas an inference is an interpretation of an observation. Students should also understand that current scientific knowledge and the perspectives of individual scientists guide both observations and inferences. Thus, different scientists can have different but equally valid interpretations of the same observations due to differences in their perspectives and background knowledge.

- *The difference between data and evidence in science:* Data are measurements, observations, and findings from other studies that are collected as part of an investigation. Evidence, in contrast, is analyzed data and an interpretation of the analysis.

Hints for Implementing the Lab

- Have your students consider why dry peas and plastic beads are included in the materials. These materials are included so that students can have a positive and negative control to conduct an actual experiment. You may need to discuss the design of controlled experiments with them.

- Consider soaking the peas overnight before the activity. You can drain them and keep them moist by covering with wet paper towels.

- Only one kind of sensor is necessary to measure respiration activity. If using O_2 sensors, the O_2 levels should decrease in the germinating peas. With CO_2 sensors, the CO_2 levels should increase with the germinating peas.

- If you do not have access to sensors, consider using a color change system involving acid-base indicators (e.g., methylene blue) and putting the peas in water.

Topic Connections

Table 1.2 (p. 34) provides an overview of the scientific practices, crosscutting concepts, disciplinary core ideas, and supporting ideas at the heart of this lab investigation. In addition, it lists NOS and NOSI concepts for the explicit and reflective discussion. Finally, it lists literacy and mathematics skills (*CCSS ELA* and *CCSS Mathematics*) that are addressed during the investigation.

LAB 1

TABLE 1.2

Lab 1 alignment with standards

Scientific practices	• Asking questions and defining problems • Developing and using models • Planning and carrying out investigations • Analyzing and interpreting data • Constructing explanations • Engaging in argument from evidence • Obtaining, evaluating, and communicating information
Crosscutting concepts	• Cause and effect: Mechanism and explanation • Energy and matter: Flows, cycles, and conservation
Core idea	• LS1: From molecules to organisms: Structures and processes
Supporting ideas	• Cellular respiration • Mitochondria • Oxygen and carbon dioxide • Energy from sugar
NOS and NOSI concepts	• Observations and inferences • Difference between data and evidence
Literacy connections (*CCSS ELA*)	• *Reading:* Key ideas and details, craft and structure, integration of knowledge and ideas • *Writing:* Text types and purposes, production and distribution of writing, research to build and present knowledge, range of writing • *Speaking and listening:* Comprehension and collaboration, presentation of knowledge and ideas
Mathematics connections (*CCSS Mathematics*)	• Make sense of problems and persevere in solving them • Reason abstractly and quantitatively • Construct viable arguments and critique the reasoning of others • Use appropriate tools strategically

National Science Teachers Association

Lab Handout

Lab 1. Cellular Respiration: Do Plants Use Cellular Respiration to Produce Energy?

Introduction

One characteristic of living things is they must take in nutrients and give off waste in order to survive. This is because all living tissues (which are made of cells) are constantly using energy. In animals, this energy comes from a reaction called *cellular respiration*. Cellular respiration refers to a process that occurs inside cells where sugar is used as a fuel source. This process happens in a specific location inside cells called the *mitochondrion* (the plural form is *mitochondria*). Figure L1.1 shows a drawing and an image from an electron microscope of a mitochondrion. Mitochondria are found in both plant and animal cells.

FIGURE L1.1 _____

(a) Drawing and (b) electron microscopic image of a mitochondrion

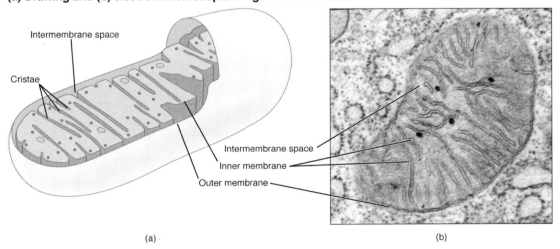

(a)   (b)

The energy that cells use comes from the chemical bonds in sugar. During cellular respiration, oxygen helps convert the chemical energy in sugar molecules into a form animals can use. The energy is transferred by moving electrons from one molecule to another. When electrons are added or taken away, new chemical bonds and types of molecules can be formed. The oxygen helps transfer electrons from sugar to the chemical bonds in another molecule. Most living organisms use a special molecule known as ATP to provide energy for all the activities taking place in their cells. The following equation describes this process:

Sugar ($C_6H_{12}O_6$) + oxygen (O_2) → water (H_2O) + carbon dioxide (CO_2) + usable energy (ATP)

We know that humans use this process to produce energy because when a human breathes, the air that he or she inhales contains about 21% O_2 and less than 1% CO_2; however, when he or she exhales, the air contains about 15% O_2 and 5% CO_2. We also know that all animals use this process to produce energy. It is a unifying characteristic of animals, but what about other types of living things like plants? Do these organisms use this process as well? In this lab investigation you will use an O_2 or CO_2 gas sensor to determine if plants use cellular respiration to produce energy just like animals.

Your Task

Design a scientific investigation to determine if plants use the process of cellular respiration to produce energy. To do this, you will need to use sensors to determine if these organisms cause a change in the CO_2 or O_2 concentrations of air.

The guiding question of this investigation is, **Do plants use cellular respiration to produce energy?**

Materials

You may use any of the following materials during your investigation:

Consumables	Equipment
• Germinating peas (i.e., peas that have been soaked in water) • Dry peas • Plastic beads	• CO_2 or O_2 gas sensor • Biochambers or sealed containers with opening for sensors • Go!Link adaptor and laptop computer • Sanitized indirectly vented chemical-splash goggles • Chemical-resistant apron • Gloves

Safety Precautions

Follow all normal lab safety rules. In addition, take the following safety precautions:

1. Put on sanitized indirectly vented chemical-splash goggles and laboratory apron and gloves before starting the lab activity.

2. Wash hands with soap and water after completing the lab activity.

Investigation Proposal Required?   ☐ Yes   ☐ No

Getting Started

During your investigation you will need to determine if the peas are producing CO_2 or using O_2. To do this, you can use CO_2 and O_2 gas sensors (see Figure L1.2). To answer the guiding question, you will need to design and conduct an experiment. To accomplish this task, you must first determine what type of data you need to collect, how you will collect it, and how you will analyze the data.

To determine *what type of data you need to collect*, think about the following questions:

- What information will tell you that cellular respiration is occurring in the peas?
- How will the sensors help you measure cellular respiration?
- What type of measurements or observations will you need to record during your investigation?

To determine *how you will collect your data*, think about the following questions:

- What will serve as a control (or comparison) condition?
- What types of treatment conditions will you need to set up and how will you do it?
- How often will you collect data and when will you do it?
- How will you make sure that your data are of high quality (i.e., how will you reduce error)?
- How will you keep track of the data you collect and how will you organize it?

To determine *how you will analyze the data*, think about the following questions:

- How will you determine if there is a difference between the treatment conditions and the control condition?
- What type of calculations will you need to make?
- What type of graph could you create to help make sense of your data?

Connections to Crosscutting Concepts, the Nature of Science, and the Nature of Scientific Inquiry

As you work through your investigation, be sure to think about

An O_2 or a CO_2 gas sensor can be used to measure changes in gas concentration.

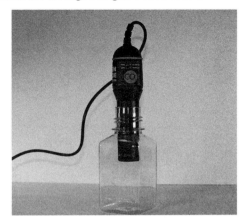

- how scientists try to figure out cause-and-effect relationships that explain why something happens,
- how energy and matter flow through living things while being totally conserved,
- how observations and inferences are different but related to each other, and
- the difference between data collected and evidence created in an investigation.

Initial Argument

Once your group has finished collecting and analyzing your data, you will need to develop an initial argument. Your argument must include a claim, evidence to support your claim, and a justification of the evidence. The claim is your group's answer to the guiding question. The evidence is an analysis and interpretation of your data. Finally, the justification of the evidence is why your group thinks the evidence matters. The justification of the evidence is important because scientists can use different kinds of evidence to support their claims. Your group will create your initial argument on a whiteboard. Your whiteboard should include all the information shown in Figure L1.3.

FIGURE L1.3 _____

Argument presentation on a whiteboard

The Guiding Question:	
Our Claim:	
Our Evidence:	Our Justification of the Evidence:

Argumentation Session

The argumentation session allows all of the groups to share their arguments. One member of each group will stay at the lab station to share that group's argument, while the other members of the group go to the other lab stations one at a time to listen to and critique the arguments developed by their classmates. This is similar to how scientists present their arguments to other scientists at conferences. If you are responsible for critiquing your classmates' arguments, your goal is to look for mistakes so these mistakes can be fixed and they can make their argument better. The argumentation session is also a good time to think about ways you can make your initial argument better. Scientists must share and critique arguments like this to develop new ideas.

To critique an argument, you might need more information than what is included on the whiteboard. You will therefore need to ask the presenter lots of questions. Here are some good questions to ask:

- What did your group do to collect the data? Why do you think that way is the best way to do it?
- What did your group do to analyze the data? Why did your group decide to analyze it that way?
- What other ways of analyzing and interpreting the data did your group talk about?

- What did your group do to make sure that these calculations are correct?

- Why did your group decide to present your evidence in that way?

- What other claims did your group discuss before you decided on that one? Why did your group abandon those other ideas?

- How sure are you that your group's claim is accurate? What could you do to be more certain?

Once the argumentation session is complete, you will have a chance to meet with your group and revise your original argument. Your group might need to gather more data or design a way to test one or more alternative claims as part of this process. Remember, your goal at this stage of the investigation is to develop the most valid or acceptable answer to the research question!

Report

Once you have completed your research, you will need to prepare an investigation report that consists of three sections that provide answers to the following questions:

1. What question were you trying to answer and why?

2. What did you do during your investigation and why did you conduct your investigation in this way?

3. What is your argument?

Your report should answer these questions in two pages or less. The report must be typed, and any diagrams, figures, or tables should be embedded into the document. Be sure to write in a persuasive style; you are trying to convince others that your claim is acceptable or valid!

LAB 1

Lab 1. Cellular Respiration: Do Plants Use Cellular Respiration to Produce Energy?

1. Susan and Jessica are having a discussion in science class about the process of cellular respiration. Susan makes the argument that animals use cellular respiration to produce energy based on the evidence that they breathe out carbon dioxide (CO_2); she also said that plants do not use cellular respiration because plants release oxygen (O_2) to the atmosphere. Jessica disagrees with Susan and claims that both plants and animals use cellular respiration. Who do you agree with, Susan or Jessica? Explain your reasoning.

2. Describe the similar structures and features of plant and animal cells that allow for cellular respiration to happen in both types of organisms.

3. In this investigation you measured the amount of gas in the biochamber with a sensor. Were those measurements data or evidence?

   a. Data

   b. Evidence

   c. Unsure

   Explain your answer.

4. Scientists always agree on their observations, but may disagree on the inferences.

   a. I agree with this statement.
   b. I disagree with this statement.

   Explain your answer, using an example from your investigation about cellular respiration.

5. An important goal in science is to develop causal explanations for observations. Explain what a causal explanation is and why it is important, using an example from your investigation about cellular respiration.

6. It is important for scientists to understand the flow of energy in a system. Explain why this is important, using an example from your investigation about cellular respiration.

LAB 2

Lab 2. Photosynthesis: Where Does Photosynthesis Take Place in Plants?

Purpose

The purpose of this lab is to *introduce* students to the process of photosynthesis. Specifically, this investigation gives students an opportunity to explore which parts of a plant demonstrate the most photosynthetic activity. Teachers will need to help students understand that carbon dioxide gas (CO_2) is used by photosynthesis and oxygen gas (O_2) is given off by it, because the concentration of these gases is how students will measure the cellular process. This lab gives students an opportunity to understand how scientific phenomena occur on different scales and with different sizes. Students will also learn how energy and matter flow through living systems in a conserved manner. Students will have the opportunity to reflect on the different methods used by scientists and the specific role that experiments play in science.

FIGURE 2.1

Major elements of photosynthesis

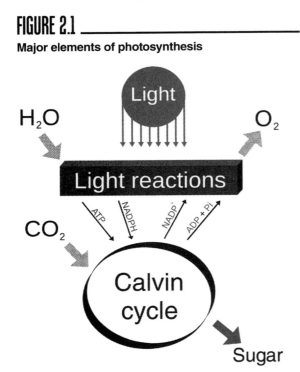

Note: ADP = adenosine diphosphate; ATP = adenosine triphosphate; $NADP^+$ = positively charged form of nicotinamide adenine dinucleotide phosphate; NADPH = nicotinamide adenine dinucleotide phosphate; Pi = inorganic phosphate.

The Content

Photosynthesis is a chemical process by which green plants produce sugar and O_2 for themselves (Figure 2.1). The sugar produced is used by a plant as food to provide energy for other activities. Animals can also eat plants to get sugar for their own energy needs. The O_2 is released into the atmosphere and is also used for other chemical reactions inside plants for producing energy.

The green found in most plants serves an important survival need. The green comes from special organelles found in plants known as *chloroplasts* (Figure 2.2), the location where photosynthesis takes place. Chloroplasts are responsible for giving plants their green color; this color comes from a chemical called *chlorophyll*, which provides green plants with a special chemical ability to absorb energy from light. That energy is used for building sugar molecules during photosynthesis.

FIGURE 2.2

Structure of a chloroplast

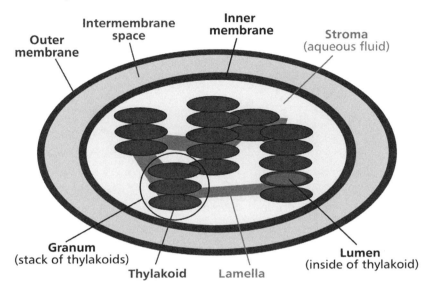

Photosynthesis requires two chemicals to react: CO_2 from the atmosphere and water from the air and the ground. The chemical equation for photosynthesis is

$$\text{Carbon dioxide } (CO_2) + \text{water } (H_2O) + \text{light energy} \rightarrow \text{sugar } (C_6H_{12}O_6) + \text{oxygen } (O_2) + \text{water } (H_2O)$$

Photosynthesis involves two major cycles of chemical reactions. One of these cycles is referred to as the *light reactions*. This name comes from the fact that during this set of reactions, light energy is captured by the chlorophyll molecules inside a plant's cells. Chlorophyll absorbs the energy from sunlight (or other light sources) and releases high-energy electrons to react with other chemicals and enzymes, while also producing O_2 as a by-product. The energy in the electrons, along with water molecules, are used to create molecules of adenosine triphosphate (ATP) and nicotinamide adenine dinucleotide phosphate (NADPH). These two molecules are used in the other reaction cycle, along with CO_2 absorbed from the atmosphere, to create sugar molecules. This reaction cycle is known as the *Calvin cycle* (named after the lead scientist who figured out all the reactions in the cycle); it is also called the dark reaction or the light-independent reactions, because light energy is not necessary for these reactions to occur. As this cycle produces sugar molecules, it also produces lower-energy versions of ATP (ADP + P) and NADPH (NADP+), which get reused and reenergized during the light reactions. Interestingly, these two different cycles of reactions do not necessarily have to occur in concert with each other. Some plants have adapted to low light and dry environments by performing these cycles during different periods of the day or under different climate conditions.

The sugar produced by photosynthesis is used to produce the flowers, leaves, stems, and roots—collectively known as the *biomass* of the plant. Thus, plants get their building blocks from air, and photosynthesis is the process that puts these building blocks together. The cells not only use the products of photosynthesis to support their own growth but also use those materials to build new cells during reproduction. All cells in a plant need to reproduce at some point, so they will all need those materials. Since cells reproduce everywhere in the plant, it would make sense that photosynthesis would need to happen everywhere, too. But does it?

Not all parts of the plant are exposed to light (the roots are not exposed), and without energy from light, photosynthesis cannot create the ATP and NADPH through the light reactions. Furthermore, to absorb the light energy, chloroplasts must be present. If chloroplasts are present, the chlorophyll in them will also give off a green color. Although chloroplasts are found mainly in plant cells, they are not found in every cell in a plant. Not all parts of a plant perform photosynthesis, and those cells that do not perform photosynthesis (i.e., root cells underground) would waste energy and material in maintaining nonfunctioning chloroplasts. However, plant cells not performing photosynthesis still have some form of *plastid*, which includes a broader group of similar organelles that serve specific functions in plant cells. Just as a chloroplast performs photosynthesis in the green leaves of a plant, amyloplasts in root cells store starch and detect gravity, which influences root growth.

Timeline

The instructional time needed to implement this lab investigation is 180–250 minutes. Appendix 2 (p. 355) provides options for implementing this lab investigation over several class periods. Option E (250 minutes) should be used if students are unfamiliar with scientific writing, because this option provides extra instructional time for scaffolding the writing process. You can scaffold the writing process by modeling, providing examples, and providing hints as students write each section of the report. Option F (180 minutes) should be used if students are familiar with scientific writing and have the skills needed to write an investigation report on their own. In option F, students complete stage 6 (writing the investigation report) and stage 8 (revising the investigation report) as homework.

Materials and Preparation

The materials needed to implement this investigation are listed in Table 2.1. You will need laptop computers to connect to and collect data from the sensors. These sensors can be ordered from a science supply company such as Vernier or Pasco.

TABLE 2.1

Materials list

Item	Quantity
Consumables	
Fresh geranium plant (can be cut into sections)	1 per group
Equipment and other materials	
Scissors	1 per group
O_2 gas sensor	1 per group
CO_2 gas sensor	1 per group
Temperature probe	1 per group
Biochambers (or sealed containers that can accommodate gas sensors)	4 per group
Go!Link adaptor and laptop computer	1 per group
Sanitized indirectly vented chemical-splash goggles	1 per student
Chemical-resistant apron	1 per student
Gloves	1 pair per student
Investigation Proposal A (optional)	1 per group
Whiteboard, 2' × 3'*	1 per group
Lab Handout	1 per student
Peer-review guide	1 per student
Checkout Questions	1 per student

* As an alternative, students can use computer and presentation software such as Microsoft PowerPoint or Apple Keynote to create their arguments.

Safety Precautions

Follow all normal lab safety rules. In addition, take the following safety precautions:

1. Put on sanitized indirectly vented chemical-splash goggles and laboratory apron and gloves before starting the lab activity.

2. Be careful when using scissors to cut up pieces of the geranium. Also, although geraniums are not very toxic for humans, do not eat any part of the plant.

3. Wash hands with soap and water after completing the lab activity.

LAB 2

Topics for the Explicit and Reflective Discussion

Concepts That Can Be Used to Justify the Evidence

To provide an adequate justification of their evidence, students must explain why they included the evidence in their arguments and make the assumptions underlying their analysis and interpretation of the data explicit. In this investigation, students can use the following concepts to help justify their evidence:

- Photosynthesis involves the trapping of light energy.
- Chloroplasts contain chlorophyll, which absorbs energy from light.
- Chlorophyll gives plants areas of green color.

We recommend that you review these concepts during the explicit and reflective discussion to help students make this connection.

How to Design Better Investigations

It is important for students to reflect on the strengths and weaknesses of the investigation they designed during the explicit and reflective discussion. Students should therefore be encouraged to discuss ways to eliminate potential flaws, measurement errors, or sources of bias in their investigations. To help students be more reflective about the design of their investigation, you can ask the following questions:

- What were some of the strengths of your investigation? What made it scientific?
- What were some of the weaknesses of your investigation? What made it less scientific?
- If you were to do this investigation again, what would you do to address the weaknesses in your investigation? What could you do to make it more scientific?

Crosscutting Concepts

This investigation is aligned with two crosscutting concepts found in *A Framework for K–12 Science Education*, and you should review these concepts during the explicit and reflective discussion.

- *Scale, proportion, and quantity:* It is critical for scientists to be able to recognize what is relevant at different sizes, time frames, and scales. Scientists must also be able to recognize proportional relationships between categories or quantities.
- *Energy and matter: Flows, cycles, and conservation:* In science it is important to track how energy and matter move into, out of, and within systems.

The Nature of Science and the Nature of Scientific Inquiry

This investigation is aligned with two important concepts related to the *nature of science* (NOS) and the *nature of scientific inquiry* (NOSI), and you should review these concepts during the explicit and reflective discussion.

- *Methods used in scientific investigations:* Examples of methods include experiments, systematic observations of a phenomenon, literature reviews, and analysis of existing data sets; the choice of method depends on the objectives of the research. There is no universal step-by-step scientific method that all scientists follow; rather, different scientific disciplines (e.g., chemistry vs. biology) and fields within a discipline (e.g., cellular biology vs. evolutionary biology) use different types of methods, use different core theories, and rely on different standards to develop scientific knowledge.

- *The nature and role of experiments:* Scientists use experiments to test the validity of a hypothesis (i.e., a tentative explanation) for an observed phenomenon. Experiments include a test and the formulation of predictions (expected results) if the test is conducted and the hypothesis is valid. The experiment is then carried out and the predictions are compared with the observed results of the experiment. If the predictions match the observed results, then the hypothesis is supported. If the predictions do not match the observed results, then the hypothesis is not supported. A signature feature of an experiment is the control of variables to help eliminate alternative explanations for observed results.

Hints for Implementing the Lab

- You can either allow students to cut up their own geraniums or have them already prepared for the students. If you allow the students to cut them, you may want to demonstrate the proper places to cut to isolate a flower, a leaf, a stem, and a root.

- Only one kind of sensor is necessary to measure respiration activity. If using CO_2 gas sensors, the CO_2 levels should decrease where photosynthesis is actively occurring. With O_2 gas sensors, the O_2 levels should increase with photosynthesis, but probably not as much.

- Remind students that they will want to create a control/comparison test that should involve an empty biochamber.

Topic Connections

Table 2.2 (p. 48) provides an overview of the scientific practices, crosscutting concepts, disciplinary core ideas, and supporting ideas at the heart of this lab investigation. In addition, it lists NOS and NOSI concepts for the explicit and reflective discussion. Finally, it lists literacy and mathematics skills (*CCSS ELA* and *CCSS Mathematics*) that are addressed during the investigation.

LAB 2

TABLE 2.2

Lab 2 alignment with standards

Scientific practices	• Asking questions and defining problems • Developing and using models • Planning and carrying out investigations • Analyzing and interpreting data • Constructing explanations • Engaging in argument from evidence • Obtaining, evaluating, and communicating information
Crosscutting concepts	• Scale, proportion, and quantity • Energy and matter: Flows, cycles, and conservation
Core idea	• LS1: From molecules to organisms: Structures and processes
Supporting ideas	• Photosynthesis • Chloroplasts • Chlorophyll • Light energy
NOS and NOSI concepts	• Methods used in scientific investigations • Nature and role of experiments
Literacy connections (*CCSS ELA*)	• *Reading:* Key ideas and details, craft and structure, integration of knowledge and ideas • *Writing:* Text types and purposes, production and distribution of writing, research to build and present knowledge, range of writing • *Speaking and listening:* Comprehension and collaboration, presentation of knowledge and ideas
Mathematics connections (*CCSS Mathematics*)	• Make sense of problems and persevere in solving them • Reason abstractly and quantitatively • Construct viable arguments and critique the reasoning of others • Use appropriate tools strategically

Lab Handout

Lab 2. Photosynthesis: Where Does Photosynthesis Take Place in Plants?

Introduction

Photosynthesis is a chemical process in which green plants produce sugar and oxygen gas (O_2) for themselves. The sugar produced is used by a plant as food to provide energy for other activities. Animals can also eat plants to get sugar for their own energy needs. The O_2 is released into the atmosphere and is also used for other chemical reactions inside plants for producing energy.

The green found in most plants serves an important survival need. The green comes from special organelles found in plants known as *chloroplasts*, which are the location where certain chemical reactions take place that help plants live. Chloroplasts are responsible for giving plants their green color; this color comes from a chemical called *chlorophyll*, which provides green plants with a special chemical ability to absorb energy from light. That energy is used for building sugar molecules during photosynthesis. Photosynthesis requires two chemicals to react: carbon dioxide (CO_2) from the atmosphere and water that can come from the ground (Figure L2.1). The chemical equation for photosynthesis is

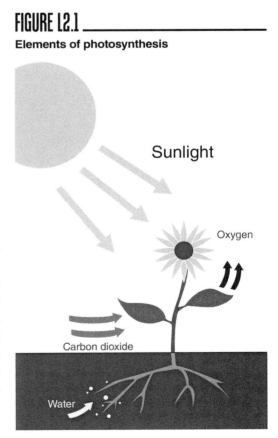

FIGURE L2.1
Elements of photosynthesis

Sunlight

Oxygen

Carbon dioxide

Water

Carbon dioxide (CO_2) + water (H_2O) + light energy
→ sugar ($C_6H_{12}O_6$) + oxygen (O_2) + water (H_2O)

This sugar is then used to produce the flowers, leaves, stems, and roots—the *biomass* of the plant. In other words, plants get their building blocks from air! Photosynthesis is the process that plants use to put these building blocks together. The cells not only use the products of photosynthesis to support their own growth but also use those materials to build new cells during reproduction. All cells in a plant need to reproduce at some point, so they will all need those materials. Since cells reproduce everywhere in the plant, it would make sense that photosynthesis would need to happen everywhere, too. But does it?

LAB 2

Your Task

Design a scientific investigation to determine if photosynthesis occurs in all of the main parts of a plant. These parts include the flowers, leaves, stems, and roots of a plant.

The guiding question of this investigation is, **Where does photosynthesis take place in plants?**

Materials

You may use any of the following materials during your investigation:

Consumables	Equipment
• Fresh geranium plant	• Scissors • O_2 gas sensor • CO_2 gas sensor • Biochambers or sealed containers with opening for sensors • Go!Link adaptor and laptop computer • Sanitized indirectly vented chemical-splash goggles • Chemical-resistant apron • Gloves

Safety Precautions

Follow all normal lab safety rules. In addition, take the following safety precautions:

1. Put on sanitized indirectly vented chemical-splash goggles and laboratory apron and gloves before starting the lab activity.

2. Be careful when using scissors to cut up pieces of the geranium. Also, although geraniums are not very toxic for humans, do not eat any part of the plant.

3. Wash hands with soap and water after completing the lab activity.

Investigation Proposal Required?   ☐ Yes   ☐ No

Getting Started

Figure L2.2 shows how CO_2 and O_2 gas sensors can be inserted into a biochamber. The sensors can then be connected to a laptop to collect data about CO_2 and O_2 gas concentration over periods of time. Ask your teacher for help if you do not understand how to set up the sensors and computer to collect data.

To answer the guiding question, you will need to design and conduct an investigation that explores rates of photosynthesis in different temperatures. To accomplish this task, you must first determine what type of data you need to collect, how you will collect it, and how you will analyze it. To determine *what type of data you need to collect*, think about the following questions:

- How will you divide the geranium plant up to test for photosynthesis?
- What data will show you that photosynthesis is occurring?
- What type of measurements or observations will you need to record during your investigation?

To determine *how you will collect your data*, think about the following questions:

- What will serve as a control (or comparison) condition?
- What types of treatment conditions will you need to set up and how will you do it?
- How often will you collect data and when will you do it?
- How will you make sure that your data are of high quality (i.e., how will you reduce error)?
- How will you keep track of the data you collect and how will you organize it?

To determine *how you will analyze your data*, think about the following questions:

- How will you determine if there is a difference between the treatment conditions and the control condition?
- What type of calculations will you need to make?
- What type of graph could you create to help make sense of your data?

FIGURE L2.2

CO_2 gas sensor

Connections to Crosscutting Concepts, the Nature of Science, and the Nature of Scientific Inquiry

As you work through your investigation, be sure to think about

- how science explores events that happen at different sizes and on different scales,
- how energy and matter flow through living things while being totally conserved,
- how science uses different methods to investigate the natural world, and
- how experiments serve a certain role in science.

LAB 2

Initial Argument

Once your group has finished collecting and analyzing your data, you will need to develop an initial argument. Your argument must include a claim, evidence to support your claim, and a justification of the evidence. The claim is your group's answer to the guiding question. The evidence is an analysis and interpretation of your data. Finally, the justification of the evidence is why your group thinks the evidence matters. The justification of the evidence is important because scientists can use different kinds of evidence to support their claims. Your group will create your initial argument on a whiteboard. Your whiteboard should include all the information shown in Figure L2.3.

FIGURE L2.3

Argument presentation on a whiteboard

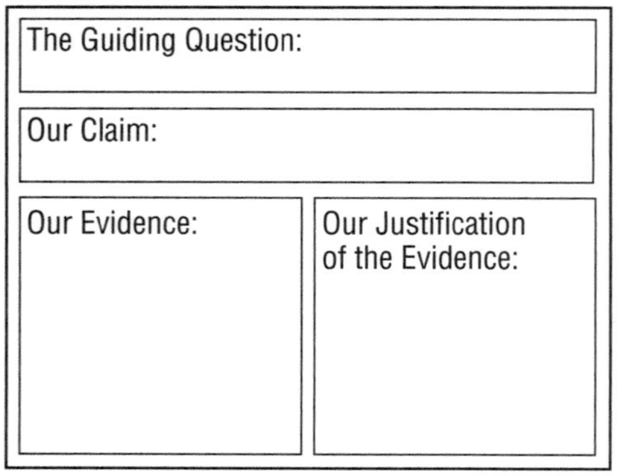

The Guiding Question:	
Our Claim:	
Our Evidence:	Our Justification of the Evidence:

Argumentation Session

The argumentation session allows all of the groups to share their arguments. One member of each group will stay at the lab station to share that group's argument, while the other members of the group go to the other lab stations one at a time to listen to and critique the arguments developed by their classmates. This is similar to how scientists present their arguments to other scientists at conferences. If you are responsible for critiquing your classmates' arguments, your goal is to look for mistakes so these mistakes can be fixed and they can make their argument better. The argumentation session is also a good time to think about ways you can make your initial argument better. Scientists must share and critique arguments like this to develop new ideas.

To critique an argument, you might need more information than what is included on the whiteboard. You will therefore need to ask the presenter lots of questions. Here are some good questions to ask:

- What did your group do to collect the data? Why do you think that way is the best way to do it?
- What did your group do to analyze the data? Why did your group decide to analyze it that way?
- What other ways of analyzing and interpreting the data did your group talk about?
- What did your group do to make sure that these calculations are correct?
- Why did your group decide to present your evidence in that way?
- What other claims did your group discuss before you decided on that one? Why did your group abandon those other ideas?
- How sure are you that your group's claim is accurate? What could you do to be more certain?

Once the argumentation session is complete, you will have a chance to meet with your group and revise your original argument. Your group might need to gather more data or design a way to test one or more alternative claims as part of this process. Remember, your goal at this stage of the investigation is to develop the most valid or acceptable answer to the research question!

Report

Once you have completed your research, you will need to prepare an investigation report that consists of three sections that provide answers to the following questions:

1. What question were you trying to answer and why?

2. What did you do during your investigation and why did you conduct your investigation in this way?

3. What is your argument?

Your report should answer these questions in two pages or less. The report must be typed, and any diagrams, figures, or tables should be embedded into the document. Be sure to write in a persuasive style; you are trying to convince others that your claim is acceptable or valid!

LAB 2

Lab 2. Photosynthesis: Where Does Photosynthesis Take Place in Plants?

1. Describe the process of photosynthesis and complete the diagram below.

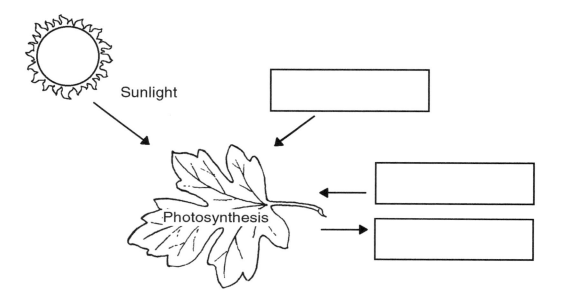

2. Two students were discussing where photosynthesis happens within a plant. One student claimed that photosynthesis only occurs in the leaves of a plant. Another student claimed that photosynthesis occurs in the leaves and flowers of a plant. The table below contains data collected during an investigation. Each container was exposed to the same amount of light for the same amount of time.

Container	Contents	Initial CO_2 level	CO_2 level after 5 hours	CO_2 level after 10 hours
A	3 leaves	900 ppm	700 ppm	500 ppm
B	7 leaves	900 ppm	500 ppm	200 ppm
C	3 leaves and 1 flower	900 ppm	750 ppm	600 ppm
D	2 flowers	900 ppm	950 ppm	1,000 ppm

Use the data above to help support the student you most agree with.

3. The data generated in the scenario for question 2 came from an experiment.

   a. I agree with this statement.

   b. I disagree with this statement.

   Explain your answer.

4. In science, experiments are better than systematic observations.

    a. I agree with this statement.

    b. I disagree with this statement.

    Explain your answer, using an example from your investigation about photosynthesis.

5. It is important for scientists to understand the relationship between different quantities and how those relationships change over time. Explain why understanding a relationship between variables is important in science by using an example from your investigation about photosynthesis.

6. It is important for scientists to understand the flow of energy in a system. Explain why this is important, using an example from your investigation about photosynthesis.

Application Labs

LAB 3

Lab 3. Osmosis: How Does the Concentration of Salt in Water Affect the Rate of Osmosis?

Purpose

The purpose of this lab is for students to *apply* their knowledge of the process of osmosis. Specifically, this investigation gives students an opportunity to examine the relationship between the concentration of a solute (salt) and the rate of osmosis. Teachers will need to help students consider how to set up controls for this experiment. This lab gives students an opportunity to understand how living systems are studied through the development of system models and how matter and energy flow through living systems. Students will also have the opportunity to reflect on the difference between observations and inferences and the specific role that experiments play in science.

The Content

In both plants and animals, each cell is surrounded by a membrane. This membrane forms a selective barrier between the cell and its environment (see Figure 3.1—the membrane is the wall in the middle of the figure). Large molecules, such as sugars ($C_6H_{12}O_6$) or fats, and charged molecules, such as sodium ions (Na^+) or chlorine ions (Cl^-), cannot pass through the membrane, but small molecules such as oxygen (O_2) can. The membrane is made up of *phospholipids* that have a polar end and a nonpolar end. The phospholipids create two layers, with the nonpolar sides of each layer facing toward the interior of the membrane and the polar sides facing out. The nonpolar interior limits the ability of charged particles to cross the membrane. Without this barrier, the substances necessary to the life of the cell would diffuse uniformly into the cell's surroundings, and toxic materials from the surroundings would enter the cell. The cell membrane is referred to as *semipermeable* because some particles can naturally cross it while others cannot. This ability to regulate the flow of molecules into and out of the cell keeps the cell's internal environment stable, even though parts of that environment are always shifting.

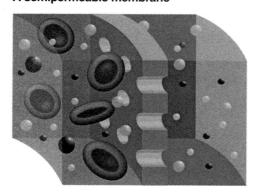

FIGURE 3.1

A semipermeable membrane

Chemical particles are constantly in motion. This motion generates from vibrations caused by the constant movement of electrons in atoms. This phenomenon is called *Brownian motion*. How much they move is related to the amount of energy they contain and how concentrated they are. *Diffusion* is the movement of chemical particles (i.e., atoms, molecules, ions) from an area of high concentration to an area of low concentration. Differences in areas of concentration create a *concentration gradient,* which represents a difference in potential energy of the particles involved. The movement of particles from high to low concentration follows a natural tendency for chemical systems to reach equilibrium. Without any barriers to such movement (like a membrane), chemical particles naturally diffuse in this direction. If a membrane is present, then only particles that can cross it naturally will be able to continue to diffuse normally. To make particles move in the opposite direction (low concentration to high concentration), energy must be added to the particles. This energy adds potential energy to the moving particles by converting the kinetic energy necessary to make the move.

Osmosis refers specifically to the diffusion of water molecules. In cells, water cannot simply diffuse across the membrane. However, special openings in the membrane allow for easy flow of water molecules so cells can take in or get rid of water when needed. These openings involve proteins that are embedded in the membrane and keep a channel open for water molecules to cross. Those proteins are known as *aquaporins.*

An *isotonic solution* is a solution that has the same concentration of particles and water as the cell. If blood cells (or other cells) are placed in contact with an isotonic solution, they will neither shrink nor swell. If the solution is *hypertonic*—having a higher concentration of solute (and lower concentration of water) than inside the cell membrane—the cells will lose water and shrink; water moves from high to low concentration in attempts to reach equilibrium. If the solution is *hypotonic*—having a lower concentration of solute and higher concentration of water molecules—the cells will gain water and swell. Saltwater from the ocean is hypertonic to the cells of the human body since it has more salt in it. Cells, as a result, lose water and shrink (Figure 3.2). That is why we can't drink water from the ocean—it dehydrates body tissues instead of quenching thirst. Cells will give off extra amounts of water to attempt to bring the body into osmotic balance.

FIGURE 3.2

(a) Red blood cells in saltwater solution and (b) normal red blood cells

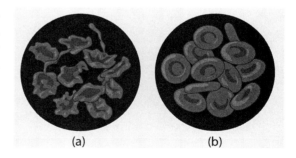

(a)   (b)

Timeline

The instructional time needed to implement this lab investigation is 180–250 minutes. Appendix 2 (p. 355) provides options for implementing this lab investigation over several class periods. Option E (250 minutes) should be used if students are unfamiliar with scientific writing, because this option provides extra instructional time for scaffolding the writing process. You can scaffold the writing process by modeling, providing examples, and providing hints as students write each section of the report. Option F (180 minutes)

LAB 3

should be used if students are familiar with scientific writing and have the skills needed to write an investigation report on their own. In option F, students complete stage 6 (writing the investigation report) and stage 8 (revising the investigation report) as homework.

Materials and Preparation

The materials needed to implement this investigation are listed in Table 3.1. Dialysis tubing is available from several chemical supply companies, including Flinn Scientific and Ward's Science. The tubing will need to be soaked for at least 20 minutes before being used.

TABLE 3.1 _____

Materials list

Item	Quantity
Consumables	
Salt solutions (high, medium, and low concentration)	50 ml each per group
Water	500 ml per group
Equipment and other materials	
Electronic or triple beam balance	1 per group
Graduated cylinder	1 per group
Beakers	3 per group
Dialysis tubing (soaked strips)	3–4 per group
Sanitized indirectly vented chemical-splash goggles	1 per student
Chemical-resistant apron	1 per student
Gloves	1 pair per student
Investigation Proposal A (optional)	1 per group
Whiteboard, 2' × 3'*	1 per group
Lab Handout	1 per student
Peer-review guide	1 per student
Checkout Questions	1 per student

* As an alternative, students can use computer and presentation software such as Microsoft PowerPoint or Apple Keynote to create their arguments.

Safety Precautions

Follow all normal lab safety rules. In addition, take the following safety precautions:

1. Put on sanitized indirectly vented chemical-splash goggles and laboratory apron and gloves before starting the lab activity.

2. Immediately wipe up any spilled water to avoid a slip and fall hazard.

3. Wash hands with soap and water after completing the lab activity.

Topics for the Explicit and Reflective Discussion

Concepts That Can Be Used to Justify the Evidence

To provide an adequate justification of their evidence, students must explain why they included the evidence in their arguments and make the assumptions underlying their analysis and interpretation of the data explicit. In this investigation, students can use the following concepts to help justify their evidence:

- In osmosis, water molecules move from high concentration to low concentration.
- The concentration of solute in a system is connected to the concentration of water in that system.
- Moving from high to low concentration naturally occurs without adding energy to the system.
- Energy is neither created nor destroyed; it only changes form (converting potential energy in concentration to kinetic energy that moves molecules along a concentration gradient).

We recommend that you review these concepts during the explicit and reflective discussion to help students make this connection.

How to Design Better Investigations

It is important for students to reflect on the strengths and weaknesses of the investigation they designed during the explicit and reflective discussion. Students should therefore be encouraged to discuss ways to eliminate potential flaws, measurement errors, or sources of bias in their investigations. To help students be more reflective about the design of their investigation, you can ask the following questions:

- What were some of the strengths of your investigation? What made it scientific?
- What were some of the weaknesses of your investigation? What made it less scientific?

- If you were to do this investigation again, what would you do to address the weaknesses in your investigation? What could you do to make it more scientific?

Crosscutting Concepts

This investigation is aligned with two crosscutting concepts found in *A Framework for K–12 Science Education,* and you should review these concepts during the explicit and reflective discussion.

- *Systems and system models:* It is critical for scientists to be able to define the system under study (e.g., organelles, cells, tissues) and then make a model of it to understand it. Models can be physical, conceptual, or mathematical.

- *Energy and matter: Flows, cycles, and conservation:* Students should realize that in science it is important to track how energy and matter move into, out of, and within systems.

The Nature of Science and the Nature of Scientific Inquiry

This investigation is aligned with two important concepts related to the *nature of science* (NOS) and the *nature of scientific inquiry* (NOSI), and you should review these concepts during the explicit and reflective discussion.

- *The difference between observations and inferences in science:* An observation is a descriptive statement about a natural phenomenon, whereas an inference is an interpretation of an observation. Students should also understand that current scientific knowledge and the perspectives of individual scientists guide both observations and inferences. Thus, different scientists can have different but equally valid interpretations of the same observations due to differences in their perspectives and background knowledge.

- *The nature and role of experiments in science:* Scientists use experiments to test the validity of a hypothesis (i.e., a tentative explanation) for an observed phenomenon. Experiments include a test and the formulation of predictions (expected results) if the test is conducted and the hypothesis is valid. The experiment is then carried out and the predictions are compared with the observed results of the experiment. If the predictions match the observed results, then the hypothesis is supported. If the predictions do not match the observed results, then the hypothesis is not supported. A signature feature of an experiment is the control of variables to help eliminate alternative explanations for observed results.

Hints for Implementing the Lab

- Demonstrate for students how to create a "cell" using the dialysis tubing during the stage 1 introduction.

- Encourage some groups to try the salt solutions inside the tubing and others to try putting the salt solutions on the outside.
- Remind students to be mindful of cross-contamination of salt solutions as they set up their experiment.
- Demonstrate how to use an electronic balance, including how to tare the balance, if they have not used one before.

Topic Connections

Table 3.2 (p. 64) provides an overview of the scientific practices, crosscutting concepts, disciplinary core ideas, and supporting ideas at the heart of this lab investigation. In addition, it lists NOS and NOSI concepts for the explicit and reflective discussion. Finally, it lists literacy and mathematics skills (*CCSS ELA* and *CCSS Mathematics*) that are addressed during the investigation.

TABLE 3.2

Lab 3 alignment with standards

Scientific practices	• Asking questions and defining problems • Developing and using models • Planning and carrying out investigations • Analyzing and interpreting data • Using mathematics and computational thinking • Constructing explanations • Engaging in argument from evidence • Obtaining, evaluating, and communicating information
Crosscutting concepts	• Systems and system models • Energy and matter: Flows, cycles, and conservation
Core idea	• LS1: From molecules to organisms: Structures and processes
Supporting ideas	• Osmosis • Concentration gradient • Conservation of energy
NOS and NOSI concepts	• Observations and inferences • Nature and role of experiments
Literacy connections (CCSS ELA)	• *Reading:* Key ideas and details, craft and structure, integration of knowledge and ideas • *Writing:* Text types and purposes, production and distribution of writing, research to build and present knowledge, range of writing • *Speaking and listening:* Comprehension and collaboration, presentation of knowledge and ideas
Mathematics connections (CCSS Mathematics)	• Make sense of problems and persevere in solving them • Reason abstractly and quantitatively • Construct viable arguments and critique the reasoning of others • Use appropriate tools strategically

Lab Handout

Lab 3. Osmosis: How Does the Concentration of Salt in Water Affect the Rate of Osmosis?

Introduction

In both plants and animals, each cell is surrounded by a membrane. This membrane forms a selective barrier between the cell and its environment (see Figure L3.1—the membrane is the wall in the middle of the figure). Large molecules, such as sugars ($C_6H_{12}O_6$) or fats, and charged molecules, such as sodium ions (Na^+) or chlorine ions (Cl^-), cannot pass through the membrane, but small molecules such as oxygen (O_2) can. Without this barrier, the substances necessary to the life of the cell would diffuse uniformly into the cell's surroundings, and toxic materials from the surroundings would enter the cell. The cell membrane is referred to as *semipermeable* because some particles can naturally cross it while others cannot. This ability to regulate the flow of molecules into and out of the cell keeps the cell's internal environment stable, even though parts of that environment are always shifting.

A semipermeable membrane

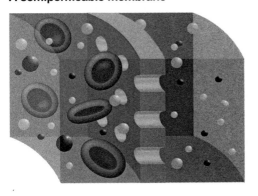

Chemical particles are constantly in motion. How much they move is related to the amount of energy they contain and how concentrated they are. *Diffusion* is the movement of chemical particles (i.e., atoms, molecules, ions) from an area of high concentration to an area of low concentration. Without any barriers to such movement (like a membrane), chemical particles naturally diffuse in this direction. If a membrane is present, then only particles that can cross it naturally will be able to continue to diffuse normally. To make particles move in the opposite direction (low concentration to high concentration), energy must be added to the particles. *Osmosis* refers specifically to the diffusion of water molecules. In cells, water cannot simply diffuse across the membrane. However, special openings in the membrane allow for easy flow of water molecules so cells can take in or get rid of water when needed.

An *isotonic* solution is a solution that has the same concentration of particles and water as the cell. If blood cells (or other cells) are placed in contact with an isotonic solution, they will neither shrink nor swell. If the solution is *hypertonic*—having a higher concentration of solute (and lower concentration of water) than inside the cell membrane—the cells will lose water and shrink. If the solution is *hypotonic*—having a lower concentration of solute and higher concentration of water molecules—the cells will gain water and swell. Saltwater from the ocean is hypertonic to the cells

LAB 3

(a) Red blood cells in saltwater solution and (b) normal red blood cells

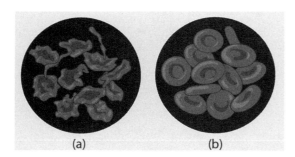

(a)   (b)

of the human body since it has more salt in it. Cells, as a result, lose water and shrink (see Figure L3.2). That is why we can't drink water from the ocean—it dehydrates body tissues instead of quenching thirst.

Your Task

Design an experiment to determine how the concentration of salt in water affects the rate of osmosis.

The guiding question of this investigation is, **How does the concentration of salt in water affect the rate of osmosis?**

Materials

You may use any of the following materials during your investigation:

Consumables	Equipment
• Salt solutions • Water	• Electronic or triple beam balance • Graduated cylinder and beakers • Dialysis tubing (assume that it behaves just like the membrane of a cell) • Sanitized indirectly vented chemical-splash goggles • Chemical-resistant apron • Gloves

Safety Precautions

Follow all normal lab safety rules. In addition, take the following safety precautions:

1. Put on sanitized indirectly vented chemical-splash goggles and laboratory apron and gloves before starting the lab activity.

2. Immediately wipe up any spilled water to avoid a slip and fall hazard.

3. Wash hands with soap and water after completing the lab activity.

Investigation Proposal Required?   ☐ Yes   ☐ No

Getting Started

You will use models of cells rather than real cells during your experiment. You will use models for two reasons: (1) a model of a cell is much larger than a real cell, which makes the process of data collection much easier; and (2) you can create your cell models in any way you see fit, which makes it easier to control for a wide range of variables during your experiment.

You can construct a model cell by using the dialysis tubing. Dialysis tubing behaves much like a cell membrane. To create a model of a cell, place the dialysis tubing in water until it is thoroughly soaked. Remove the soaked tubing from the water and tightly twist one end several times and either tie with string or tie a knot in the tubing. You can then fill the model cell with a salt solution or distilled water. Once filled, twist the open end several times and tie it tightly as shown in Figure L3.3. You can then dry the bag and place it into any type of solution you need.

FIGURE L3.3

Tying the dialysis tubing

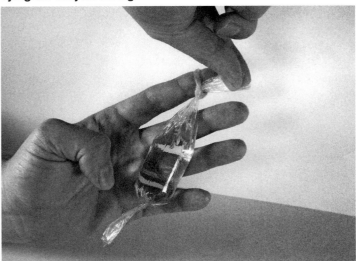

To answer the guiding question, you will need to design and conduct an experiment. To accomplish this task, you must first determine what type of data you need to collect, how you will collect it, and how you will analyze it before you can design your experiment. To determine *what type of data you need to collect*, think about the following questions:

- How will you determine the rate of osmosis?
- What type of measurements or observations will you need to record during your investigation?

To determine *how you will collect your data*, think about the following questions:

- What will serve as a control (or comparison) condition?
- What types of treatment conditions will you need to set up and how will you do it?
- How often will you collect data and when will you do it?
- How will you make sure that your data are of high quality (i.e., how will you reduce error)?
- How will you keep track of the data you collect and how will you organize it?

To determine *how you will analyze your data*, think about the following questions:

- How will you determine if there is a difference between the treatment conditions and the control condition?
- How will you calculate change over time?
- What type of graph could you create to help make sense of your data?

Connections to Crosscutting Concepts, the Nature of Science, and the Nature of Scientific Inquiry

As you work through your investigation, be sure to think about

- why developing and using models is important in science,
- the importance of tracking how matter flows into and out of a system,
- the difference between observations and inferences in science, and
- the nature and role of experiments in science.

Initial Argument

Once your group has finished collecting and analyzing your data, you will need to develop an initial argument. Your argument must include a claim, evidence to support your claim, and a justification of the evidence. The claim is your group's answer to the guiding question. The evidence is an analysis and interpretation of your data. Finally, the justification of the evidence is why your group thinks the evidence matters. The justification of the evidence is important because scientists can use different kinds of evidence to support their claims. Your group will create your initial argument on a whiteboard. Your whiteboard should include all the information shown in Figure L3.4.

FIGURE L3.4 _____

Argument presentation on a whiteboard

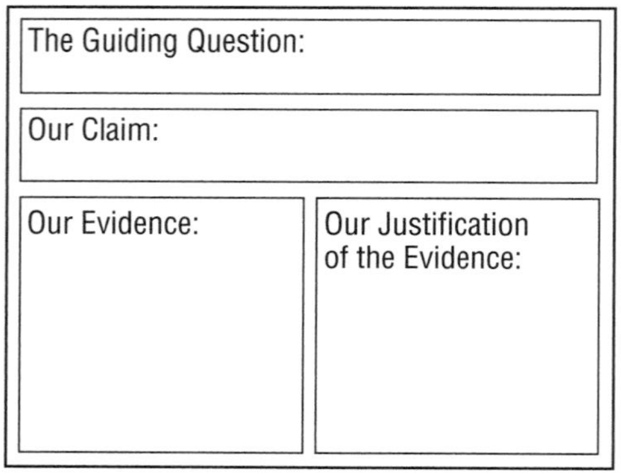

The Guiding Question:	
Our Claim:	
Our Evidence:	Our Justification of the Evidence:

Argumentation Session

The argumentation session allows all of the groups to share their arguments. One member of each group will stay at the lab station to share that group's argument, while the other members of the group go to the other lab stations one at a time to listen to and critique the arguments developed by their classmates. This is similar to how scientists present their arguments to other scientists at conferences. If you are responsible for critiquing your classmates' arguments, your goal is to look for mistakes so these mistakes can be fixed and they can make their argument better. The argumentation session is also a good time to think about ways you can make your initial argument better. Scientists must share and critique arguments like this to develop new ideas.

To critique an argument, you might need more information than what is included on the whiteboard. You will therefore need to ask the presenter lots of questions. Here are some good questions to ask:

- What did your group do to collect the data? Why do you think that way is the best way to do it?

- What did your group do to analyze the data? Why did your group decide to analyze it that way?

- What other ways of analyzing and interpreting the data did your group talk about?

- What did your group do to make sure that these calculations are correct?

- Why did your group decide to present your evidence in that way?

- What other claims did your group discuss before you decided on that one? Why did your group abandon those other ideas?

- How sure are you that your group's claim is accurate? What could you do to be more certain?

Once the argumentation session is complete, you will have a chance to meet with your group and revise your initial argument. Your group might need to gather more data or design a way to test one or more alternative claims as part of this process. Remember, your goal at this stage of the investigation is to develop the most valid or acceptable answer to the research question!

Report

Once you have completed your research, you will need to prepare an investigation report that consists of three sections that provide answers to the following questions:

1. What question were you trying to answer and why?

2. What did you do during your investigation and why did you conduct your investigation in this way?

3. What is your argument?

Your report should answer these questions in two pages or less. The report must be typed, and any diagrams, figures, or tables should be embedded into the document. Be sure to write in a persuasive style; you are trying to convince others that your claim is acceptable or valid!

LAB 3

Lab 3. Osmosis: How Does the Concentration of Salt in Water Affect the Rate of Osmosis?

1. Describe the process of osmosis.

2. A potato was cut into 10 equal-size cubes, each weighing about 10 grams. The cubes were placed into five different beakers of saltwater, each with a different concentration (%) of salt solution.

The potatoes were allowed to sit in the salt solution for 24 hours and then removed from the beakers, dried, and weighed. The figure below shows the average percent change in mass for the potatoes.

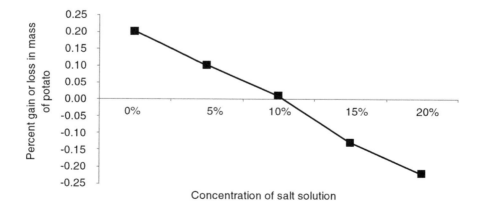

Using what you know about osmosis and the data above, what was the original concentration of salt solution in the potato? Explain your reasoning.

3. The data generated in the scenario for question 2 above came from an experiment.

   a. I agree with this statement.
   b. I disagree with this statement.

   Explain your answer, using an example from your investigation about osmosis.

4. In this investigation we observed osmosis happening.

   a. I agree with this statement.
   b. I disagree with this statement.

   Explain your answer, using an example from your investigation about osmosis.

5. It is important for scientists to develop and use models. Explain why models are important in science by using an example from your investigation about osmosis.

6. It is important for scientists to understand the flow of matter in a system. Explain why this is important, using an example from your investigation about osmosis.

Teacher Notes

Lab 4. Cell Structure: What Type of Cell Is on the Unknown Slides?

Purpose

The purpose of this lab is for students to *apply* their understanding of cell structure to categorize whether unknown cells come from plants or animals. Specifically, this investigation gives students an opportunity to explore the components of cell structure visible with a compound light microscope and use those observations to analyze unknown samples. Teachers will need to help students understand how to use and adjust a microscope, including changing magnification and adjusting image resolution. This lab gives students an opportunity to learn how scientists look for patterns during their work and how the structure of living things affects their functions. Students will also have the opportunity to reflect on the difference between observations and inferences and how scientific knowledge changes over time.

The Content

Scientists who study living organisms deal with a lot of different types of life forms, from trees to tadpoles and bacteria to birds. As they investigate how life happens on the planet, they rely on several scientific theories that have developed over time. These theories combine different types of evidence to support a big idea that explains some aspect of life or the natural world. One of the major theories that scientists rely on when studying living things is the *basic cell theory*. This theory includes three major ideas that have been supported over the years as new life forms continue to be discovered:

1. All living organisms are made up of one or more cells.

2. The cell is the basic unit of life.

3. All new cells come from cells that are already alive.

The tenets of the basic cell theory are some of the most important ideas needed to understand how living things grow, function, reproduce, and change over time. By understanding how the basic unit of the cell works, scientists have been able to develop a fuller understanding of how organisms are structured and how they interact with their environments.

The development of the cell theory coincided with the rapidly expanding creation of new technologies. In the 1600s, Robert Hooke was the first scientist to describe the structure of a cell. The name "cell" came from Hooke's observations of cork tissue, where he noted that the repeating structure and boxlike shape was reminiscent of a monk's cell, or room. Following Hooke, scientists continued to explore the cellular structures found in a variety of organisms. In the mid-1800s, several scientists (Robert Remak, Matthias Schleiden, Theodor Schwann, and Rudolf Virchow) developed different lines of evidence that would be used to create the basic cell theory. Continued scientific investigations have led to increasingly more complex understandings of how the cell is structured and how it functions. Thus, the modern cell theory includes several ideas that go beyond the basic cell theory, including the following ideas:

- Although cells can function independently or in smaller groups, the overall activity of an organism depends on the combined activity of all its cells.
- Cells are similar in chemical composition, including several biochemical reactions and cycles (like cellular respiration), across species.
- Cells contain DNA and RNA, the primary genetic material for life.

Just as there are many types of organisms, including plants and animals, there are also many types of cells. However, there are several features found in all cells. The most common features are the presence of DNA and the presence of a cell membrane. DNA is a molecule that contains information that cells need to live. The cell membrane is the sheet of molecules that separates the inside of the cell from the rest of the environment. More complex cells, like those found in animals and plants, have other structures in common, known as *organelles*. Organelles are special structures found inside cells that serve different functions. Those functions include helping the cell get energy, making the materials it needs to continue growing, and storing the information (like DNA) to make new cells.

Plant and animal cells have many organelles in common, including the nucleus (involved in storing genetic material and directing cellular activity), the endoplasmic reticulum (involved in construction of necessary molecules), Golgi bodies (involved in packaging and delivery of cellular materials), ribosomes (involved in protein construction), the cell membrane (involved in regulation of intake and output of materials), and mitochondria (involved in cellular respiration for energy demands). Figure 4.1 (p. 74) illustrates some of these organelles.

Some organelles found in plant cells, however, are not found in animal cells and vice versa. For example, animal cells have centrioles (which help organize cell division in animal cells), but plant cells do not. Plant cells have an extra layer surrounding them called a cell wall. Cell walls are stiff membranes that sit outside of the cell membrane and help keep plant cells in a specific shape. The differences in types of organelles can be used to distinguish between cells that come from a plant and cells that come from an animal. However, not all organelles can be seen using microscopes we use in school.

Another organelle commonly associated with plant cells is a *chloroplast*. Chloroplasts are enclosed, green organelles that are the site of photosynthesis, a sugar-producing chemical process. They get their green color from a pigment called chlorophyll, which is responsible

FIGURE 4.1

Animal cell diagram

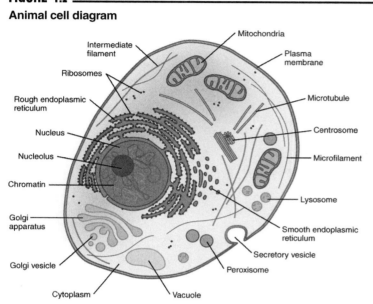

for capturing light energy for photosynthesis. Although chloroplasts are found mainly in plant cells, they are not found in every cell in a plant. Not all parts of a plant perform photosynthesis, and those cells that do not perform photosynthesis (i.e., root cells underground) would waste energy and material in maintaining nonfunctioning chloroplasts. However, plant cells not performing photosynthesis will still have some form of *plastid*, which includes a broader group of similar organelles that serve specific functions in plant cells. Just as a chloroplast performs photosynthesis in the green leaves of a plant, amyloplasts in root cells store starch and detect gravity, which influences root growth.

Timeline

The instructional time needed to implement this lab investigation is 130–200 minutes. Appendix 2 (p. 355) provides options for implementing this lab investigation over several class periods. Option C (200 minutes) should be used if students are unfamiliar with scientific writing, because this option provides extra instructional time for scaffolding the writing process. You can scaffold the writing process by modeling, providing examples, and providing hints as students write each section of the report. Option D (130 minutes) should be used if students are familiar with scientific writing and have the skills needed to write an investigation report on their own. In option D, students complete stage 6 (writing the investigation report) and stage 8 (revising the investigation report) as homework.

Materials and Preparation

The materials needed to implement this investigation are listed in Table 4.1. The slides can be purchased in sets or individually from a science supply company such as Carolina or Ward's Science.

TABLE 4.1

Materials list

Item	Quantity
Known slide A (plant cells)	1 per group
Known slide B (plant cells)	1 per group
Known slide C (animal cells)	1 per group
Known slide D (animal cells)	1 per group
Unknown slide E	1 per group
Unknown slide F	1 per group
Unknown slide G	1 per group
Unknown slide H	1 per group
Compound light microscope	1 per group
Slide wipes	1 pack per group
Sanitized indirectly vented chemical-splash goggles	1 per student
Chemical-resistant apron	1 per student
Gloves	1 pair per student
Investigation Proposal A (optional)	1 per group
Whiteboard, 2' × 3'*	1 per group
Lab Handout	1 per student
Peer-review guide	1 per student
Checkout Questions	1 per student

* As an alternative, students can use computer and presentation software such as Microsoft PowerPoint or Apple Keynote to create their arguments.

Safety Precautions

Follow all normal lab safety rules. In addition, take the following safety precautions:

1. Put on sanitized indirectly vented chemical-splash goggles and laboratory apron and gloves before starting the lab activity.

2. Handle all glassware with care to avoid breakage. Sharp edges can cut skin!

3. Follow all safety rules that apply when working with electrical equipment, and use only GFCI-protected electrical receptacles.

4. Wash hands with soap and water after completing the lab activity.

LAB 4

Topics for the Explicit and Reflective Discussion

Concepts That Can Be Used to Justify the Evidence

To provide an adequate justification of their evidence, students must explain why they included the evidence in their arguments and make the assumptions underlying their analysis and interpretation of the data explicit. In this investigation, students can use the following concepts to help justify their evidence:

- The tenets of cell theory explain how living things grow, function, reproduce, and change over time.
- Cells share many common organelle structures.
- Some organelles are unique to animals (centrioles) and some to plants (cell wall).
- Organelles present in a cell affect the function of that cell.

We recommend that you review these concepts during the explicit and reflective discussion to help students make this connection.

How to Design Better Investigations

It is important for students to reflect on the strengths and weaknesses of the investigation they designed during the explicit and reflective discussion. Students should therefore be encouraged to discuss ways to eliminate potential flaws, measurement errors, or sources of bias in their investigations. To help students be more reflective about the design of their investigation, you can ask the following questions:

- What were some of the strengths of your investigation? What made it scientific?
- What were some of the weaknesses of your investigation? What made it less scientific?
- If you were to do this investigation again, what would you do to address the weaknesses in your investigation? What could you do to make it more scientific?

Crosscutting Concepts

This investigation is aligned with two crosscutting concepts found in *A Framework for K–12 Science Education,* and you should review these concepts during the explicit and reflective discussion.

- *Patterns:* Scientists look for patterns in nature and attempt to understand the underlying cause of these patterns. Biologists, for example, look for patterns in the common and different structures in living things.
- *Structure and function:* In nature, the way a living thing is structured or shaped determines how it functions and places limits on what it can and cannot do.

The Nature of Science and the Nature of Scientific Inquiry

This investigation is aligned with two important concepts related to the *nature of science* (NOS) and the *nature of scientific inquiry* (NOSI), and you should review these concepts during the explicit and reflective discussion.

- *The differences between observations and inferences:* An observation is a descriptive statement about a natural phenomenon, whereas an inference is an interpretation of an observation. Students should also understand that current scientific knowledge and the perspectives of individual scientists guide both observations and inferences. Thus, different scientists can have different but equally valid interpretations of the same observations due to differences in their perspectives and background knowledge.

- *Changes in scientific knowledge over time:* A person can have confidence in the validity of scientific knowledge but must also accept that scientific knowledge may be abandoned or modified in light of new evidence or because existing evidence has been reconceptualized by scientists. There are many examples in the history of science of both evolutionary changes (i.e., the slow or gradual refinement of ideas) and revolutionary changes (i.e., the rapid abandonment of a well-established idea) in scientific knowledge.

Hints for Implementing the Lab

- Try to select some plant slides that do not have chloroplasts in them.
- Give your students time to learn how to use a microscope if you have not previously taught them this skill.
- Have groups record the set of criteria they use to determine the unknown slides based on their observations of the known ones.

Topic Connections

Table 4.2 (p. 78) provides an overview of the scientific practices, crosscutting concepts, disciplinary core ideas, and supporting ideas at the heart of this lab investigation. In addition, it lists NOS and NOSI concepts for the explicit and reflective discussion. Finally, it lists literacy and mathematics skills (*CCSS ELA* and *CCSS Mathematics*) that are addressed during the investigation.

TABLE 4.2

Lab 4 alignment with standards

Scientific practices	• Asking questions and defining problems • Developing and using models • Planning and carrying out investigations • Analyzing and interpreting data • Constructing explanations • Engaging in argument from evidence • Obtaining, evaluating, and communicating information
Crosscutting concepts	• Patterns • Structure and function
Core idea	• LS1: From molecules to organisms: Structures and processes
Supporting ideas	• Cell theory • Organelles • Effect of organelles on cell function
NOS and NOSI concepts	• Observations and inferences • Changes in scientific knowledge over time
Literacy connections (CCSS ELA)	• *Reading:* Key ideas and details, craft and structure, integration of knowledge and ideas • *Writing:* Text types and purposes, production and distribution of writing, research to build and present knowledge, range of writing • *Speaking and listening:* Comprehension and collaboration, presentation of knowledge and ideas
Mathematics connections (CCSS Mathematics)	• Make sense of problems and persevere in solving them • Construct viable arguments and critique the reasoning of others

Lab Handout

Lab 4. Cell Structure: What Type of Cell Is on the Unknown Slides?

Introduction

Scientists who study living organisms deal with a lot of different types of life forms, from trees to tadpoles and bacteria to birds. As they investigate how life happens on the planet, they rely on several scientific theories that have developed over time. These theories combine different types of evidence to support a big idea that explains some aspect of life or the natural world. One of the major theories that scientists rely on when studying living things is the *cell theory*. This theory includes three major ideas that have been supported over the years as new life forms continue to be discovered:

1. All living organisms are made up of one or more cells.

2. The cell is the basic unit of life.

3. All new cells come from cells that are already alive.

Just as there are many types of organisms, including plants and animals, there are also many types of cells. However, there are several features found in all cells. The most common features are the presence of DNA and the presence of a *cell membrane*. DNA is a molecule that contains information that cells need to live. The cell membrane is the sheet of molecules that separates the inside of the cell from the rest of the environment. You can think of the cell membrane as a cell's "skin." More complex cells, like those found in animals and plants, have other structures in common, known as *organelles*. Organelles are special structures found inside cells that serve different functions. Those functions include helping the cell get energy, making the materials it needs to continue growing, and storing the information (like DNA) to make new cells. The organelles present in a cell will also influence what activities that cell can perform.

Plant and animal cells have many organelles in common, including the nucleus, the endoplasmic reticulum, Golgi bodies, ribosomes, the cell membrane, and mitochondria (see Figure L4.1, p. 80). Some organelles found in plant cells, however, are not found in animal cells, and vice versa. For example, animal cells have centrioles (which help organize cell division in animal cells), but plant cells do not. Plant cells have an extra layer surrounding them called a cell wall. Cell walls are stiff membranes that sit outside of the cell membrane and help keep plant cells in a specific shape. The differences in types of organelles can be used to distinguish between cells that come from a plant and cells that come from an animal. However, not all organelles can be seen using microscopes we use in school.

LAB 4

FIGURE L4.1
Animal cell diagram

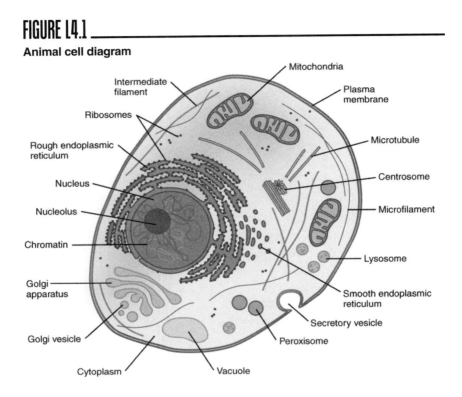

Your Task

Using what you learn from observations of several slides of cells using a microscope, with some slides labeled as plant cells and others as animal cells, determine what types of cells are on the slides labeled as "unknown."

The guiding question of this investigation is, **What type of cell is on the unknown slides?**

Materials

You may use any of the following materials during your investigation:

- Known slide A (plant cells)
- Known slide B (plant cells)
- Known slide C (animal cells)
- Known slide D (animal cells)
- Unknown slide E
- Unknown slide F
- Unknown slide G

- Unknown slide H
- Compound light microscope
- Slide wipes
- Sanitized indirectly vented chemical-splash goggles
- Chemical-resistant apron
- Gloves

Safety Precautions

Follow all normal lab safety rules. In addition, take the following safety precautions:

1. Put on sanitized indirectly vented chemical-splash goggles and laboratory apron and gloves before starting the lab activity.

2. Handle all glassware with care to avoid breakage. Sharp edges can cut skin!

3. Follow all safety rules that apply when working with electrical equipment, and use only GFCI-protected electrical receptacles.

4. Wash hands with soap and water after completing the lab activity.

Investigation Proposal Required?   ☐ Yes   ☐ No

Getting Started

To determine the difference between a plant cell and animal cell, you and your group will need to explore what cell structures you can see with a compound light microscope. To answer the guiding question, you must first determine what type of data you need to collect, how you will collect it, and how you will analyze it. To determine *what type of data you need to collect*, think about the following questions:

- What type of measurements or observations will you need to make during your investigation?
- How will you quantify any differences or similarities you observe in the different cells?

To determine *how you will collect your data*, think about the following questions:

- How will you make sure that your data are of high quality (i.e., how will you reduce error)?
- How will you keep track of the data you collect and how will you organize it?

To determine *how you will analyze your data*, think about the following question:

- How will you define the different categories of cells (e.g., what makes a plant cell a plant cell, what makes an animal cell an animal cell)?

Connections to Crosscutting Concepts, the Nature of Science, the Nature of Scientific Inquiry

As you work through your investigation, be sure to think about

- how scientists look for patterns across different living things,

LAB 4

- how the structure of an organelle or cell is related to the function it performs,
- the difference between observations and inferences, and
- how science knowledge changes over time as new evidence is discovered and technology is created.

Initial Argument

Once your group has finished collecting and analyzing your data, you will need to develop an initial argument. Your argument must include a claim, evidence to support your claim, and a justification of the evidence. The claim is your group's answer to the guiding question. The evidence is an analysis and interpretation of your data. Finally, the justification of the evidence is why your group thinks the evidence matters. The justification of the evidence is important because scientists can use different kinds of evidence to support their claims. Your group will create your initial argument on a whiteboard. Your whiteboard should include all the information shown in Figure L4.2.

FIGURE L4.2

Argument presentation on a whiteboard

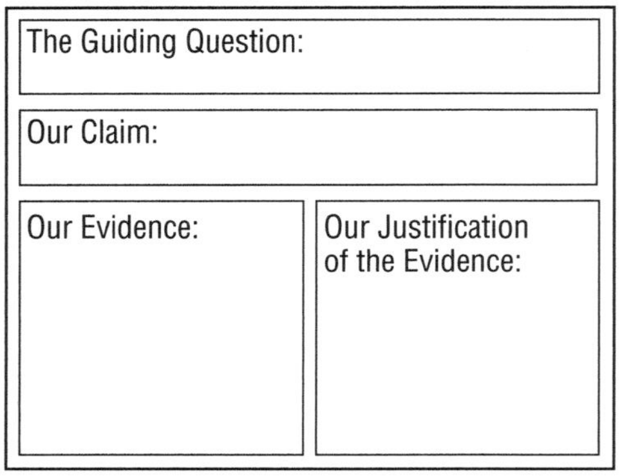

The Guiding Question:	
Our Claim:	
Our Evidence:	Our Justification of the Evidence:

Argumentation Session

The argumentation session allows all of the groups to share their arguments. One member of each group will stay at the lab station to share that group's argument, while the other members of the group go to the other lab stations one at a time to listen to and critique the arguments developed by their classmates. This is similar to how scientists present their arguments to other scientists at conferences. If you are responsible for critiquing your classmates' arguments, your goal is to look for mistakes so these mistakes can be fixed and they can make their argument better. The argumentation session is also a good time to think about ways you can make your initial argument better. Scientists must share and critique arguments like this to develop new ideas.

To critique an argument, you might need more information than what is included on the whiteboard. You will therefore need to ask the presenter lots of questions. Here are some good questions to ask:

- What did your group do to collect the data? Why do you think that way is the best way to do it?
- What did your group do to analyze the data? Why did your group decide to analyze it that way?
- What other ways of analyzing and interpreting the data did your group talk about?
- Why did your group decide to present your evidence in that way?

- What other claims did your group discuss before you decided on that one? Why did your group abandon those other ideas?

- How sure are you that your group's claim is accurate? What could you do to be more certain?

Once the argumentation session is complete, you will have a chance to meet with your group and revise your original argument. Your group might need to gather more data or design a way to test one or more alternative claims as part of this process. Remember, your goal at this stage of the investigation is to develop the most valid or acceptable answer to the research question!

Report

Once you have completed your research, you will need to prepare an investigation report that consists of three sections that provide answers to the following questions:

1. What question were you trying to answer and why?

2. What did you do during your investigation and why did you conduct your investigation in this way?

3. What is your argument?

Your report should answer these questions in two pages or less. This report must be typed, and any diagrams, figures, or tables should be embedded into the document. Be sure to write in a persuasive style; you are trying to convince others that your claim is acceptable or valid!

LAB 4

Lab 4. Cell Structure: What Type of Cell Is on the Unknown Slides?

1. Describe the characteristics that plant cells and animal cells have in common.

2. Describe the features that allow you to distinguish between plant cells and animal cells.

3. In science, it is not possible to make an inference without first observing.

    a. I agree with this statement.
    b. I disagree with this statement.

    Explain your answer, using an example from your investigation about cell structure.

4. Once a scientific idea is developed, it does not change.

    a. I agree with this statement.

    b. I disagree with this statement.

    Explain your answer, using an example from your investigation about cell structure.

5. It is important for scientists to look for and identify patterns in nature. Explain why identifying patterns is useful in science by using an example from your investigation about cell structure.

6. It is important for scientists to understand the relationship between the structure of an organism and its function. Explain why this is important, using an example from your investigation about cell structure.

LAB 5

Teacher Notes

Lab 5. Temperature and Photosynthesis: How Does Temperature Affect the Rate of Photosynthesis in Plants?

Purpose

The purpose of this lab is for students to *apply* their knowledge of photosynthesis and how to measure the chemical process. Specifically, this investigation gives students an opportunity to explore how changes in temperature influence photosynthesis. Teachers will need to help students consider how they will measure the rate of photosynthesis over an extended period of time and at different temperatures. This lab gives students an opportunity to understand how different events and effects on a system can be connected to specific causes. Students will also learn how energy and matter interact in living systems. Students will have the opportunity to reflect on the difference between data and evidence and on the different methods used by scientists.

FIGURE 5.1 _____

Major elements of photosynthesis

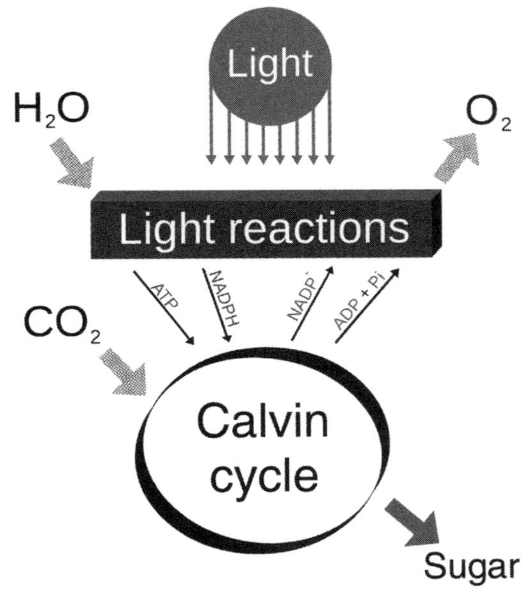

Note: ADP = adenosine diphosphate; ATP = adenosine triphosphate; NADP⁺ = positively charged form of nicotinamide adenine dinucleotide phosphate; NADPH = nicotinamide adenine dinucleotide phosphate; Pi = inorganic phosphate.

The Content

Green is the most common color among plant life on the planet. The green found in most plants actually serves an important survival need. The color comes from organelles known as *chloroplasts*. Chloroplasts contain *chlorophyll*, a green pigment that captures light energy that is used during the process of photosynthesis.

Photosynthesis is a complex chemical process by which green plants produce sugar and oxygen gas (O_2) for themselves (Figure 5.1). Photosynthesis requires two chemicals to react: carbon dioxide gas (CO_2) from the atmosphere and water from the air and the ground. However, photosynthesis also requires light energy to make sugar and oxygen. The chemical equation for photosynthesis is

Carbon dioxide (CO_2) + water (H_2O) + light energy
→ sugar ($C_6H_{12}O_6$) + oxygen (O_2) + water (H_2O)

Photosynthesis involves two major cycles of chemical reactions. One of these cycles is referred to as the *light reactions*. This name comes from the fact that during this set of reactions, light energy is captured by the chlorophyll molecules inside a plant's cells. Chlorophyll absorbs the energy from sunlight (or other light sources) and releases high-energy electrons to react with other chemicals and enzymes, while also producing O_2 as a by-product. The energy in the electrons, along with water molecules, is used to create molecules of adenosine triphosphate (ATP) and nicotinamide adenine dinucleotide phosphate (NADPH). These two molecules are used in the other reaction cycle, along with CO_2 absorbed from the atmosphere, to create sugar molecules. This reaction cycle is known as the *Calvin cycle* (named after the lead scientist who figured out all the reactions in the cycle); it is also called the dark reaction or the light-independent reactions, because light energy is not necessary for these reactions to occur. As this cycle produces sugar molecules, it also produces lower-energy versions of ATP (ADP + P) and NADPH (NADP+) which get reused and reenergized during the light reactions.

The sugar produced by photosynthesis is used by a plant as food to provide energy for other activities. This sugar is also used to produce the flowers, leaves, stems, and roots—collectively known as the *biomass* of the plant. Animals can eat plants to get sugar for their own energy needs. The O_2 is released into the atmosphere and is also used for other chemical reactions inside plants and animals for producing energy. This common process for producing energy in living things is known as *respiration*, which refers to a series of events and chemical reactions that use O_2 to break bonds in the sugar molecules, releasing the energy stored there. That energy is used to produce ATP molecules that the plants and animals can use to fuel other reactions and functions in their bodies. The products of respiration include CO_2, which is released into the atmosphere and can then be used in photosynthesis. The two processes of photosynthesis and respiration are key components of a larger cycle that supports life in many ecosystems (see Figure 5.2).

FIGURE 5.2

Interaction of photosynthesis and cellular respiration through different kinds of organisms

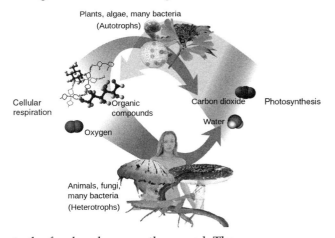

Plants rely on gases in the air and sunlight to create the food and energy they need. The amount of sunlight in a plant's environment will influence how much photosynthesis can occur. The amount of water present (dry desert vs. moist rainforest) also influences plants' activities and affects their growth. Obviously, plant growth and survival depend in part on conditions in the environment. A common example of this is the difference between the seasons. Most plants do not show a lot of activity during the winter, although they remain alive. In the spring, they become colorful and grow a lot. That growth continues over the summer, with some plants producing fruits and vegetables, and then their growth slows down during the fall. This leads to the question, If overall plant growth and behavior are affected by seasonal changes in temperature, how do different temperatures affect the process of photosynthesis? The full answer involves both photosynthesis and respiration.

As temperature increases, the rate of respiration in many organisms also increases. Within plants, this means that its cells will need more of the reactants necessary for respiration, including sugar and O_2. The easiest sources for those reactants are the products of photosynthesis. Thus, as respiration increases in plants, so does photosynthesis. For plants to grow, the rate of photosynthesis must be higher than the rate of respiration. This makes sense if you see photosynthesis as providing the building materials for plants and respiration as consuming those materials. However, photosynthesis also requires light energy, which is only available during certain periods of the day, which also happen to be when the temperature is warmer as well. Extending this idea further, the greatest amount of light and warmer temperatures is typically available during the spring and summer, which are also the periods of the greatest plant growth.

Although you should see a similar relationship between temperature and rate of photosynthesis in your students' data, they may need to consider the impact of other environmental factors on plants, including the season in which they are doing the investigation. The type of plant also can be an important factor, because some plants are hardier in certain kinds of environmental conditions (e.g., cactus plant in the desert, marigold during winter).

Timeline

The instructional time needed to implement this lab investigation is 180–250 minutes. Appendix 2 (p. 355) provides options for implementing this lab investigation over several class periods. Option E (250 minutes) should be used if students are unfamiliar with scientific writing, because this option provides extra instructional time for scaffolding the writing process. You can scaffold the writing process by modeling, providing examples, and providing hints as students write each section of the report. Option F (180 minutes) should be used if students are familiar with scientific writing and have the skills needed to write an investigation report on their own. In option F, students complete stage 6 (writing the investigation report) and stage 8 (revising the investigation report) as homework.

Materials and Preparation

The materials needed to implement this investigation are listed in Table 5.1. You will need laptop computers to connect to and collect data from the sensors. These sensors can be ordered from a science supply company such as Vernier or Pasco. If you do not have both sensors available, then use the one you have and ask students consider if a decrease or increase in the gas indicates photosynthetic activity.

TABLE 5.1
Materials list

Item	Quantity
Consumable	
Fresh geranium plant	1 per group
Equipment and other materials	
O_2 gas sensor	1 per group
CO_2 gas sensor	1 per group
Temperature probe	1 per group
Biochamber (or sealed container that can accommodate gas sensors)	1 per group
Go!Link adaptor and laptop computer	1 per group
Hot plate (or incubator)	1 per group
Ice packs (or ice bath)	4 per group (or 1 per group)
Light source (flood lamp)	1 per group
Sanitized indirectly vented chemical-splash goggles	1 per student
Chemical-resistant apron	1 per student
Gloves	1 pair per student
Investigation Proposal A (optional)	1 per group
Whiteboard, 2' × 3'*	1 per group
Lab Handout	1 per student
Peer-review guide	1 per student
Checkout Questions	1 per student

* As an alternative, students can use computer and presentation software such as Microsoft PowerPoint or Apple Keynote to create their arguments.

LAB 5

Safety Precautions

Follow all normal lab safety rules. In addition, take the following safety precautions:

1. Put on sanitized indirectly vented chemical-splash goggles and laboratory apron and gloves before starting the lab activity.

2. Be careful handling hot plates and incubators set at high temperatures because they may be hot enough to burn you.

3. Wash hands with soap and water after completing the lab activity.

Topics for the Explicit and Reflective Discussion

Concepts That Can Be Used to Justify the Evidence

To provide an adequate justification of their evidence, students must explain why they included the evidence in their arguments and make the assumptions underlying their analysis and interpretation of the data explicit. In this investigation, students can use the following concepts to help justify their evidence:

- Photosynthesis involves the use of CO_2 to produce O_2.
- Photosynthesis requires light energy to work, which in nature can also be related to the average temperature.
- Photosynthesis and respiration are two cycles that use each other's products as their reactants.

We recommend that you review these concepts during the explicit and reflective discussion to help students make this connection.

How to Design Better Investigations

It is important for students to reflect on the strengths and weaknesses of the investigation they designed during the explicit and reflective discussion. Students should therefore be encouraged to discuss ways to eliminate potential flaws, measurement errors, or sources of bias in their investigations. To help students be more reflective about the design of their investigation, you can ask the following questions:

- What were some of the strengths of your investigation? What made it scientific?
- What were some of the weaknesses of your investigation? What made it less scientific?
- If you were to do this investigation again, what would you do to address the weaknesses in your investigation? What could you do to make it more scientific?

Crosscutting Concepts

This investigation is aligned with two crosscutting concepts found in *A Framework for K–12 Science Education,* and you should review these concepts during the explicit and reflective discussion.

- *Cause and effect: Mechanism and explanation:* Natural phenomena have causes, and uncovering causal relationships (e.g., how changes in environmental conditions influence the size of cells) is a major activity of science.
- *Energy and matter: Flow, cycles, and conservation:* In science it is important to track how energy and matter move into, out of, and within systems.

The Nature of Science and the Nature of Scientific Inquiry

This investigation is aligned with two important concepts related to the *nature of science* (NOS) and the *nature of scientific inquiry* (NOSI), and you should review these concepts during the explicit and reflective discussion.

- *The difference between laws and theories in science:* A scientific law describes the behavior of a natural phenomenon or a generalized relationship under certain conditions; a scientific theory is a well-substantiated explanation of some aspect of the natural world. Theories do not become laws even with additional evidence; they explain laws. However, not all scientific laws have an accompanying explanatory theory. It is also important for students to understand that scientists do not discover laws or theories; the scientific community develops them over time.
- *The difference between data and evidence in science:* Data are measurements, observations, and findings from other studies that are collected as part of an investigation. Evidence, in contrast, is analyzed data and an interpretation of the analysis.

Hints for Implementing the Lab

- Check with a local florist or garden shop manager to see if there is a plant better suited to this activity with respect to your local environment.
- Demonstrate how the data collection software works for the various probes and sensors used.
- Emphasize to students that they will need to focus on photosynthesis over a period of time, not just one or two quick measurements.
- Remind students to consider the effect of heat from the light source on the temperature of the system. They may need to consider how close they position the light source and if they need to add some form of heat shield.

LAB 5

Topic Connections

Table 5.2 provides an overview of the scientific practices, crosscutting concepts, disciplinary core ideas, and supporting ideas at the heart of this lab investigation. In addition, it lists NOS and NOSI concepts for the explicit and reflective discussion. Finally, it lists literacy and mathematics skills (*CCSS ELA* and *CCSS Mathematics*) that are addressed during the investigation.

TABLE 5.2

Lab 5 alignment with standards

Scientific practices	• Asking questions and defining problems • Developing and using models • Planning and carrying out investigations • Analyzing and interpreting data • Using mathematics and computational thinking • Constructing explanations • Engaging in argument from evidence • Obtaining, evaluating, and communicating information
Crosscutting concepts	• Cause and effect: Mechanism and explanation • Energy and matter: Flows, cycles, and conservation
Core idea	• LS1: From molecules to organisms: Structures and processes
Supporting ideas	• Photosynthesis • Light energy • Seasonal temperatures • Respiration
NOS and NOSI concepts	• Scientific laws and theories • Difference between data and evidence
Literacy connections (*CCSS ELA*)	• *Reading:* Key ideas and details, craft and structure, integration of knowledge and ideas • *Writing:* Text types and purposes, production and distribution of writing, research to build and present knowledge, range of writing • *Speaking and listening:* Comprehension and collaboration, presentation of knowledge and ideas
Mathematics connections (*CCSS Mathematics*)	• Make sense of problems and persevere in solving them • Reason abstractly and quantitatively • Construct viable arguments and critique the reasoning of others • Model with mathematics • Use appropriate tools strategically • Attend to precision • Look for and make use of structure • Look for and express regularity in repeated reasoning

Lab Handout

Lab 5. Temperature and Photosynthesis: How Does Temperature Affect the Rate of Photosynthesis in Plants?

Introduction

All the colors that our eyes can see can be found in plants that live in the world. They include deep, bright reds to rich blues and purples. Interestingly, one common color that most plants share is green. This common characteristic is not simply due to chance. The green found in most plants actually serves an important survival need. The green comes from special organelles found in plants known as *chloroplasts*. Chloroplasts are the location where certain chemical reactions take place that help plants live. Chloroplasts are responsible for giving plants their green color; this color comes from a chemical called chlorophyll, which provides green plants with a special chemical ability to absorb energy from light. Green plants have the ability to produce their own supply of sugar through the process of *photosynthesis*.

Photosynthesis is a complex chemical process in which green plants produce sugar and oxygen gas (O_2) for themselves. The sugar produced is used by a plant as food to provide energy for other activities. Animals can also eat plants to get sugar for their own energy needs. The O_2 is released into the atmosphere (see Figure L5.1) and is used for other chemical reactions inside plants for producing energy. Animals also use O_2 to produce energy. This common process for producing energy in living things is known as *respiration*, which refers to a series of chemical reactions that use O_2 to break bonds in the sugar molecules, releasing the energy stored there. The products of respiration include carbon dioxide gas (CO_2), which is released into the atmosphere and can then be used in photosynthesis.

Photosynthesis requires two chemicals to react: CO_2 from the atmosphere and water that can come from the ground. However, photosynthesis also requires light energy to make sugar and O_2. The chemical equation for photosynthesis is

Carbon dioxide (CO_2) + water (H_2O) + light energy
→ sugar ($C_6H_{12}O_6$) + oxygen (O_2) + water (H_2O)

FIGURE L5.1 _____

Elements of photosynthesis

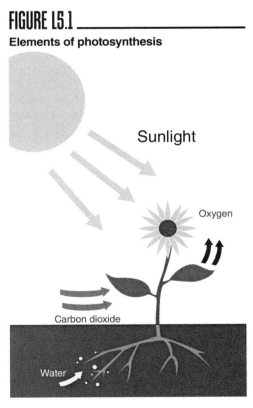

This sugar is then used to produce the flowers, leaves, stems, and roots—the *biomass* of the plant. In other words, plants get their building blocks from air!

Plants rely on gases in the air and sunlight to create the food and energy they need. The amount of sunlight in a plant's environment will influence how much photosynthesis can occur. The amount of water present (dry desert vs. moist rainforest) also influences plants' activities and affects their growth. Obviously, plant growth and survival depend in part on conditions in the environment. A common example of this is the difference between the seasons. Most plants do not show a lot of activity during the winter, although they remain alive. In the spring, they become colorful and grow a lot. That growth continues over the summer, with some plants producing fruits and vegetables, and then their growth slows down during the fall. We also know that the temperature in the environment is very different across the seasons. So if overall plant growth and behavior are affected by seasonal changes in temperature, how do different temperatures affect the process of photosynthesis?

Your Task

Design a scientific investigation to determine how temperature influences the rate of photosynthesis in plants.

The guiding question of this investigation is, **How does temperature affect the rate of photosynthesis in plants?**

Materials

You may use any of the following materials during your investigation:

Consumables	Equipment
• Fresh geranium plant	• O_2 gas sensor • CO_2 gas sensor • Temperature probe • Go!Link adaptor and laptop computer • Biochamber or sealed container with opening for sensors • Hot plate or incubator • Ice packs or ice bath • Sanitized indirectly vented chemical-splash goggles • Chemical-resistant apron • Gloves

Safety Precautions

Follow all normal lab safety rules. In addition, take the following safety precautions:

1. Put on sanitized indirectly vented chemical-splash goggles and laboratory apron and gloves before starting the lab activity.

2. Be careful handling hot plates and incubators set at high temperatures because they may be hot enough to burn you.

3. Wash hands with soap and water after completing the lab activity.

Investigation Proposal Required? ☐ Yes ☐ No

Getting Started

Figure L5.2 shows how CO_2 and O_2 gas sensors can be inserted into a biochamber. The sensors can then be connected to a laptop to collect data about CO_2 and O_2 gas concentration over periods of time. Ask your teacher for help if you do not understand how to set up the sensors and computer to collect data.

FIGURE L5.2 _____

CO_2 gas sensor

To answer the guiding question, you will need to design and conduct an investigation that explores rates of photosynthesis in different temperatures. To accomplish this task, you must determine what type of data you need to collect, how you will collect it, and how you will analyze it. To determine *what type of data you need to collect*, think about the following questions:

- How will you test the effect of temperature on the photosynthesis rate?
- How will you change the temperature inside the chamber?
- What type of measurements or observations will you need to record during your investigation?

To determine *how you will collect your data*, think about the following questions:

- What will serve as a control (or comparison) condition?
- What types of treatment conditions will you need to set up and how will you do it?
- How often will you collect data and when will you do it?
- How will you make sure that your data are of high quality (i.e., how will you reduce error)?
- How will you keep track of the data you collect and how will you organize it?

To determine *how you will analyze your data*, think about the following questions:

- How will you determine if there is a difference between the treatment conditions and the control condition?
- What type of calculations will you need to make?
- What type of graph could you create to help make sense of your data?

Connections to Crosscutting Concepts, the Nature of Science, and the Nature of Scientific Inquiry

As you work through your investigation, be sure to think about

- how scientists try to figure out cause-and-effect relationships that explain why something happens,
- how energy and matter flow through living things while being totally conserved,
- the different roles theories and laws play in science, and
- the difference between data collected and evidence created in an investigation.

Initial Argument

Once your group has finished collecting and analyzing your data, you will need to develop an initial argument. Your argument must include a claim, evidence to support your claim, and a justification of the evidence. The claim is your group's answer to the guiding question. The evidence is an analysis and interpretation of your data. Finally, the justification of the evidence is why your group thinks the evidence matters. The justification of the evidence is important because scientists can use different kinds of evidence to support their claims. Your group will create your initial argument on a whiteboard. Your whiteboard should include all the information shown in Figure L5.3.

FIGURE L5.3

Argument presentation on a whiteboard

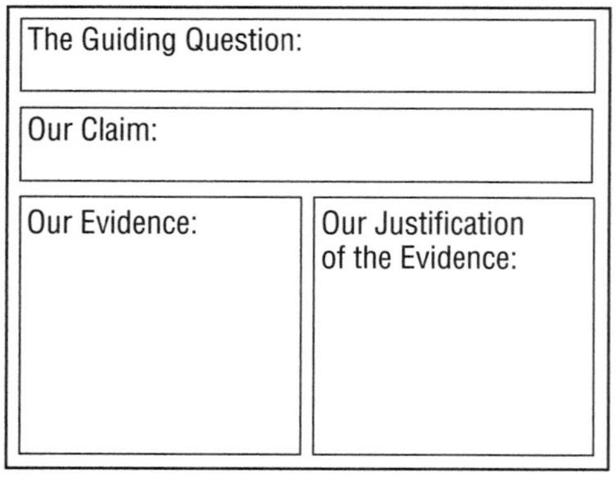

The Guiding Question:	
Our Claim:	
Our Evidence:	Our Justification of the Evidence:

Argumentation Session

The argumentation session allows all of the groups to share their arguments. One member of each group will stay at the lab station to share that group's argument, while the other members of the group go to the other lab stations one at a time to listen to and critique the arguments developed by their classmates. This is similar to how scientists present their arguments to other scientists at conferences. If you are responsible for critiquing your classmates' arguments, your goal is to look for mistakes so these mistakes can be fixed and they can make their argument better. The argumentation session is also a good time to think about ways you can make your initial argument better. Scientists must share and critique arguments like this to develop new ideas.

To critique an argument, you might need more information than what is included on the whiteboard. You will therefore need to ask the presenter lots of questions. Here are some good questions to ask:

- What did your group do to collect the data? Why do you think that way is the best way to do it?
- What did your group do to analyze the data? Why did your group decide to analyze it that way?
- What other ways of analyzing and interpreting the data did your group talk about?
- What did your group do to make sure that these calculations are correct?
- Why did your group decide to present your evidence in that way?
- What other claims did your group discuss before you decided on that one? Why did your group abandon those other ideas?
- How sure are you that your group's claim is accurate? What could you do to be more certain?

Once the argumentation session is complete, you will have a chance to meet with your group and revise your original argument. Your group might need to gather more data or design a way to test one or more alternative claims as part of this process. Remember, your goal at this stage of the investigation is to develop the most valid or acceptable answer to the research question!

Report

Once you have completed your research, you will need to prepare an investigation report that consists of three sections that provide answers to the following questions:

1. What question were you trying to answer and why?
2. What did you do during your investigation and why did you conduct your investigation in this way?
3. What is your argument?

Your report should answer these questions in two pages or less. This report must be typed, and any diagrams, figures, or tables should be embedded into the document. Be sure to write in a persuasive style; you are trying to convince others that your claim is acceptable or valid!

Checkout Questions

Lab 5. Temperature and Photosynthesis: How Does Temperature Affect the Rate of Photosynthesis in Plants?

Use the following information to answer questions 1 and 2.

Andre owns a tree farm where he grows trees and then sells them to people to plant in their yard. Andre can't sell his trees until they grow to 8 feet tall, which takes about six months. The faster the trees grow, the sooner he can sell them and make more money. Andre wanted to investigate how he could help his trees grow faster without spending money on extra fertilizer or water for the plants. He had the idea of growing a sample of trees in a greenhouse where he kept the temperature warm, about 85°F (29.4°C), and another sample in a greenhouse where he kept the temperature cooler, about 65°F (18.3°C). Andre watered all the plants the same and made sure both groups got the same amount of sunlight. After just four months the trees in the warm greenhouse had all reached 8 feet tall, but the trees in the cooler greenhouse only grew to 5 feet.

1. For a plant to grow, what must be valid about the rate of photosynthesis compared with the rate of cellular respiration? Explain your reasoning.

2. Use what you know about photosynthesis and temperature to explain why the trees in the warm greenhouse grew taller than the ones in the cool greenhouse.

3. A theory and a law serve the same purpose in science.

   a. I agree with this statement.
   b. I disagree with this statement.

Explain your answer, using an example from your investigation about temperature and photosynthesis.

4. In science, it is not possible to make an inference without first observing.

   a. I agree with this statement.
   b. I disagree with this statement.

   Explain your answer, using an example from your investigation about temperature and photosynthesis.

5. Scientists try to identify the effect that certain factors can cause in a system. Explain why identifying cause-and-effect relationships is useful in science, using an example from your investigation about temperature and photosynthesis.

6. It is important for scientists to understand the relationship between energy and matter and how they move through a system. Explain why this is important, using an example from your investigation about temperature and photosynthesis.

LAB 6

Lab 6. Energy in Food: Which Type of Nut Is Best for a New Energy Bar?

Purpose

The purpose of this lab is for students to *apply* their understanding of energy flow and transfer within a system. Specifically, this investigation gives students an opportunity to explore the relationship between the foods we eat and the energy that is provided to our bodies to carry out familiar tasks such as exercising. Student will generate arguments about which nut would be best for making an energy bar, relying on evidence such as energy in the nut, how much mass is need to provide a sufficient amount of calories, and cost of using a particular nut. This lab gives students the opportunity to understand how energy and matter flow through living systems and how the structure of organisms and molecules determines the functions they can perform. Students will also have the opportunity to reflect on the difference between data and evidence in science and the role of scientists' or engineers' imagination and creativity when solving problems.

The Content

One characteristic of living things is they must take in nutrients and give off waste in order to survive. This is because all living tissues (which are made of cells) are constantly using energy. In animals, this energy comes from a reaction called *cellular respiration*. Cellular respiration refers to a process that occurs inside cells where sugar is used as a fuel source. The process of cellular respiration is very similar to combustion reaction in chemistry. When students burn the various nuts during this lab, the matter in the nut is involved in a combustion reaction that releases a large amount of energy. The following equation describes a simplified example of the combustion reaction that takes place when the nut is burned.

Sugar ($C_6H_{12}O_6$) + oxygen (O_2) → water (H_2O) + carbon dioxide (CO_2) + usable energy

In this example sugar is used as the carbohydrate molecule that is combining with oxygen during the combustion reaction. When the various nuts are burned during this lab, the overall reaction and the energy that is released are due to the combination of burning carbohydrates, fats, proteins, and other parts of the nuts, such as the nondigestible fiber. Additionally, students will likely have heard that we "burn calories" or that we need energy to burn so that we can complete tasks. Tapping into students' notion of "burning" is helpful in this context and the combustion comparison is useful, but it does not tell the whole story with respect to the many different types of molecules that our bodies use as fuel.

Calorimetry is a process, or technique, that can be used to generate an estimate of how much chemical energy is stored in a food. Burning a food via combustion releases the energy stored in the bonds of the food molecules (carbohydrates, fats, proteins) as heat. In calorimetry the energy released by the food is used to heat up a small amount of water. One *calorie* is defined as the amount of energy needed to raise the temperature of 1 gram of water by 1°C. A *food calorie*, however, is defined as 1,000 calories (or 1 kilocalorie). Therefore, if a food label says that one serving contains 150 Calories (capital C is used when referring to food calories), there are really 150,000 calories in that serving of food. The energy in food can also be described using the unit of joules (J). One calorie is equal to 4.18 J, which is the metric unit for energy and commonly used on food labels in European countries.

Determining the actual amount of energy that is released by the burning nut and absorbed by the sample of water requires measuring the change in temperature of the water and using the following equation to solve for the amount of heat energy absorbed by the water:

$$q = m \times c \times \Delta T$$

In the equation above, q is the amount of heat (in calories, NOT food calories) absorbed by the water; m is the mass of the water sample used; c is the specific heat of water (1 calorie/g · °C); and ΔT ($T_{final} - T_{initial}$) is the change in temperature of the water sample (in °C). So what is the amount of energy (food calories) in a peanut if a 50 g sample of water has an initial temperature of 22°C and is heated to 41°C by burning the peanut? Here is a sample calculation:

If $m = 50$ g, $c = 1$ calorie/g · °C, and $\Delta T = 19°C$ (41°C – 22°C), then

$$q = (50 \text{ g}) \times (1 \text{ calorie/g} \cdot °C) \times (19°C)$$
$$q = (50 \text{ g}) \times (1 \text{ calorie/g} \cdot °C) \times (19°C)$$
$$q = 950 \text{ calories}$$

If 1 food calorie = 1,000 calories, then

$$q = (950 \text{ calories}) \times (1 \text{ food calorie/1,000 calories})$$
$$q = (950 \text{ calories}) \times (1 \text{ food calorie/1,000 calories})$$
$$q = 0.950 \text{ food calories per peanut}$$

LAB 6

It is important to note that not every peanut (or other nuts used in this investigation) has the same mass. Therefore, it is important to also consider the mass of each nut that is burned. Finding the mass of each nut before and after burning will allow the students to have a standard comparison of food calories per gram of food for each of the different nuts (pre- and post-measurements of mass are essential because not all pieces of the nut will burn to completion).

This particular activity goes beyond the science content described above and asks students to make a decision regarding which nut would be the best for making a new energy bar. Completing such a task requires that the students use a variety of evidence to support their choice for which nut is the best. Given the full context of this task, the students must make several decisions based on factors not directly measured during the investigation or decisions that are grounded in nonscientific factors, such as:

- How does the mass of the bar (total amount of ingredients) affect the price?
- Should we include more than one type of nut?
- Is the price of the bar more important than the amount of energy it provides?

Although these kinds of questions may not directly link to typical science concepts, they have important connections to economic or technological aspects that are associated with the scientific enterprise and the relationship between scientific investigations and applications of engineering principles used to solve problems.

Timeline

The instructional time needed to implement this lab investigation is 180–250 minutes. Appendix 2 (p. 355) provides options for implementing this lab investigation over several class periods. Data collection associated with this investigation can take extra time to conduct multiple trials, and options A and B both provide that extra time. Option A (250 minutes) should be used if students are unfamiliar with scientific writing, because this option provides extra instructional time for scaffolding the writing process. You can scaffold the writing process by modeling, providing examples, and providing hints as students write each section of the report. Option B (180 minutes) should be used if students are familiar with scientific writing and have the skills needed to write an investigation report on their own. In option B, students complete stage 6 (writing the investigation report) and stage 8 (revising the investigation report) as homework.

Materials and Preparation

The materials needed to implement this investigation are listed in Table 6.1. Calorimeters and combustion stands (cork stopper with a needle or paper clip for holding a nut) can be purchased from most science supply companies, but we recommend the Economy Choice

Calorimeter from Flinn Scientific. Alternatively, you can construct your own calorimeter using a basic aluminum soda can, ring stand, and thermometer (see Figure 6.1, p. 104).

TABLE 6.1

Materials list

Item	Quantity
Consumables	
Peanuts	2–5 per group
Cashews	2–5 per group
Pecans	2–5 per group
Almonds	2–5 per group
Walnuts	2–5 per group
Matches	2–5 per group
Equipment and other materials	
Calorimeter	1 per group
Combustion stand	1 per group
Thermometer or temperature sensor	1 per group
Graduated cylinder, 100 ml	1 per group
Timer	1 per group
Electronic or triple beam balance	1 per group
Sanitized indirectly vented chemical-splash goggles	1 per student
Chemical-resistant apron	1 per student
Gloves	1 pair per student
Investigation Proposal C (optional)	1 per group
Whiteboard, 2' × 3'*	1 per group
Lab Handout	1 per student
Lab 6 Reference Sheet: Costs and Exercise Calories	1 per student
Peer-review guide	1 per student
Checkout Questions	1 per student

* As an alternative, students can use computer and presentation software such as Microsoft PowerPoint or Apple Keynote to create their arguments.

LAB 6

FIGURE 6.1

Soda can calorimeter

Safety Precautions

Remind students that even though typical food items are being used during this investigation, they should treat these items like any other chemicals in the lab. Emphasize that students should not eat the food items during this investigation or remove them from the lab and eat them. In addition, tell students to take the following safety precautions:

1. Students with an allergy to nuts should notify the teacher immediately.

2. Put on sanitized indirectly vented chemical-splash goggles and laboratory apron and gloves before starting the lab activity.

3. Students should never put consumables in their mouths.

4. Use caution when working with open flames while burning the nuts. Open flames can burn skin, and combustibles and flammables must be kept away from the open flame. Students with long hair should tie it back behind their heads.

5. Handle all glassware with care to avoid breakage. Sharp glass can cut skin!

6. Wash hands with soap and water after completing the lab activity.

Topics for the Explicit and Reflective Discussion

Concepts That Can Be Used to Justify the Evidence

To provide an adequate justification of their evidence, students must explain why they included the evidence in their arguments and make the assumptions underlying their analysis and interpretation of the data explicit. For this investigation, students' justification of their evidence will likely include science concepts as well as issues related to the economics or feasibility of making an energy bar out of a specific type of nut. In this investigation, students can use the following concepts to help justify their evidence:

- Energy given off during combustion of a food (peanut, pecan, etc.) is similar to the energy released during cellular respiration.

- Energy absorbed by the water of the calorimeter is directly related to the chemical energy stored in the various nuts.

- Choosing certain nuts may result in an energy bar that is cheaper to produce.

- Depending on which nut is used, an energy bar may provide more energy, have a smaller size, or be more cost-effective based on calories per dollar.

We recommend that you review these concepts during the explicit and reflective discussion to help students make this connection.

How to Design Better Investigations

It is important for students to reflect on the strengths and weaknesses of the investigation they designed during the explicit and reflective discussion. Students should therefore be encouraged to discuss ways to eliminate potential flaws, measurement errors, or sources of bias in their investigations. To help students be more reflective about the design of their investigation, you can ask the following questions:

- What were some of the strengths of your investigation? What made it scientific?
- What were some of the weaknesses of your investigation? What made it less scientific?
- If you were to do this investigation again, what would you do to address the weaknesses in your investigation? What could you do to make it more scientific?

Crosscutting Concepts

This investigation is aligned with two crosscutting concepts found in *A Framework for K–12 Science Education,* and you should review these concepts during the explicit and reflective discussion.

- *Energy and matter: Flows, cycles, and conservation:* In science it is important to track how energy and matter move into, out of, and within systems. This lab emphasizes the chemical energy that flows from food to the consumer.
- *Structure and function:* The way an object is shaped or structured determines many of its properties and functions. Specifically, the composition of a food source dictates the amount of energy that source can provide.

The Nature of Science and the Nature of Scientific Inquiry

This investigation is aligned with two important concepts related to the *nature of science* (NOS) and the *nature of scientific inquiry* (NOSI), and you should review these concepts during the explicit and reflective discussion.

- *The difference between data and evidence in science:* Data are measurements, observations, and findings from other studies that are collected as part of an investigation. Evidence, in contrast, is analyzed data and an interpretation of the analysis.

- *The importance of imagination and creativity in science:* Students should learn that developing explanations for or models of natural phenomena and then figuring out how they can be put to the test of reality is as creative as writing poetry, composing music, or designing skyscrapers. Scientists must also use their imagination and creativity to figure out new ways to test ideas and collect or analyze data.

Hints for Implementing the Lab

- Learn how to use the calorimeter and combustion stand before the lab begins. It is important for you to know how to use the equipment so you can help students when technical issues arise.

- Allow the students to become familiar with the calorimeter as part of the tool talk before they begin to design their investigation. This gives students a chance to see what they can and cannot do with the equipment.

- It is typical that the nuts will stop burning (and the flame will go out) before the combustion process is complete. When this happens, have students relight the sample so they can continue heating the water.

- Encourage students to test all the different types of nuts available and conduct multiple trials for each. If time is an issue, consider having students work in groups of four and then have pairs within each group collect data for different nuts. This will reduce the time required to collect data, but it also increases the amount of equipment needed for the lab.

- The Lab Handout does not include a discussion of how to calculate the energy released by burning the nuts. Including a timely discussion of how to conduct those calculations will increase the rigor of this investigation. Alternatively, students can compare the mass of each nut burned with the net temperature change of the water and generate a ratio for comparison purposes. This requires that the mass of water be consistent for each trial.

Topic Connections

Table 6.2 provides an overview of the scientific practices, crosscutting concepts, disciplinary core ideas, and supporting ideas at the heart of this lab investigation. In addition, it lists NOS and NOSI concepts for the explicit and reflective discussion. Finally, it lists literacy and mathematics skills (*CCSS ELA* and *CCSS Mathematics*) that are addressed during the investigation.

TABLE 6.2 _____

Lab 6 alignment with standards

Scientific practices	• Asking questions and defining problems • Planning and carrying out investigations • Analyzing and interpreting data • Constructing explanations and designing solutions • Engaging in argument from evidence • Obtaining, evaluating, and communicating information
Crosscutting concepts	• Energy and matter: Flows, cycles, and conservation • Structure and function
Core idea	• LS1: From molecules to organisms: Structures and processes
Supporting ideas	• Cellular respiration • Energy from carbohydrates, fats, and proteins • Energy flow and transfer
NOS and NOSI concepts	• Difference between data and evidence • Imagination and creativity in science
Literacy connections (*CCSS ELA*)	• *Reading:* Key ideas and details, craft and structure, integration of knowledge and ideas • *Writing:* Text types and purposes, production and distribution of writing, research to build and present knowledge, range of writing • *Speaking and listening:* Comprehension and collaboration, presentation of knowledge and ideas
Mathematics connections (*CCSS Mathematics*)	• Make sense of problems and persevere in solving them • Reason abstractly and quantitatively • Construct viable arguments and critique the reasoning of others • Model with mathematics • Use appropriate tools strategically • Attend to precision • Look for and make use of structure • Look for and express regularity in repeated reasoning

LAB 6

Lab 6. Energy in Food: Which Type of Nut Is Best for a New Energy Bar?

Introduction

All living things must take in nutrients to survive. In plants, the process of photosynthesis converts light energy from the sun, along with carbon dioxide and water to generate sugar molecules that the plant then uses for chemical energy to complete cellular processes that are required for the plant to live and grow. Animals (including humans), however, must consume food to obtain the nutrients and chemical energy they need to complete the cellular processes that allow them to live and grow. People get energy to live by consuming a variety of food types, such as carbohydrates, proteins, and fats. Each of these food types has different properties and provides our bodies with different amounts of energy. Plants and animals break down nutrients to release chemical energy through a process called *cellular respiration*. During cellular respiration, large molecules containing carbon, hydrogen, and oxygen atoms are broken down into smaller molecules in a process similar to the combustion (or burning) of wood in a fire.

The amount of chemical energy that is stored in a food source is measured in *food calories*. A food calorie is also called a *kilocalorie* and is defined as the amount of energy needed to heat 1 kilogram of water by 1 degree Celsius. The chemical energy stored in food is released through the processes of digestion and cellular respiration so that our bodies can then use the energy for important tasks such as moving our muscles. Different food types contain different amounts of energy that our bodies can use; in general, fats provide the most energy at about 9 food calories per gram. Carbohydrates and proteins both provide about 4 food calories per gram.

Many foods contain a mix of carbohydrates, protein, and fats; this is true of nuts (peanuts, pecans, almonds, etc.). The nutrients found in different foods are shown on the *nutrition facts* label that is included on most food items sold in the United States (see Figure L6.1 for an example of this label). These labels also include information about the types of vitamins found in foods, such as vitamin A or vitamin C, or minerals found in food, such as calcium or iron. Vitamins and minerals are important for certain cellular processes, but they do not provide any calories or energy for our bodies.

Each day our bodies need a minimum number of calories so that we have enough energy to complete the tasks that keep us alive, such as maintaining our core body temperature, breathing, and keeping our heart beating and circulating blood and oxygen throughout our body. The amount of calories an individual needs to complete the most basic functions of life is called the *resting metabolic rate*. Resting metabolic rates differ from person to person based on gender, height, weight, and many other factors. People need additional energy so they can walk, talk, or do other everyday tasks, and people who exercise or play sports need even more energy so that they can complete the tasks associated with those activities such as running, jumping, or lifting heavy objects. To get all the energy we need on a daily basis, we must eat enough food to provide our bodies with all the calories we need. On average, an adult should consume about 2,400 food calories a day. In addition to eating three meals a day, sometimes it is necessary to have a snack to take in enough calories for the day. Many athletes, for example, must eat "energy bars" before they exercise or even during long exercise sessions to make sure they get the extra calories they need.

Your Task

Collect data to help you determine which type of nut would be the best to use for a new energy bar. To do this, you will need to measure the amount of energy in a variety of nuts and consider several other factors, such as cost, size of bar, and/or amount of calories that are necessary to complete different exercises.

The guiding question of this investigation is, **Which type of nut is best for a new energy bar?**

FIGURE L6.1

Example of a nutrition facts label

Nutrition Facts	
Serving Size 2/3 cup (55g)	
Servings Per Container About 8	

Amount Per Serving	
Calories 230	Calories from Fat 40

	% Daily Value*
Total Fat 8g	**12%**
Saturated Fat 1g	**5%**
Trans Fat 0g	
Cholesterol 0mg	**0%**
Sodium 160mg	**7%**
Total Carbohydrate 37g	**12%**
Dietary Fiber 4g	**16%**
Sugars 1g	
Protein 3g	

Vitamin A	10%
Vitamin C	8%
Calcium	20%
Iron	45%

* Percent Daily Values are based on a 2,000 calorie diet. Your daily value may be higher or lower depending on your calorie needs.

	Calories:	2,000	2,500
Total Fat	Less than	65g	80g
Sat Fat	Less than	20g	25g
Cholesterol	Less than	300mg	300mg
Sodium	Less than	2,400mg	2,400mg
Total Carbohydrate		300g	375g
Dietary Fiber		25g	30g

Materials

You may use any of the following materials during your investigation:

Consumables	Equipment
• Nuts (peanut, cashew, pecan, almond, walnut) • Matches	• Calorimeter • Combustion stand • Thermometer or temperature sensor • Graduated cylinder (100 ml) • Timer • Electronic or triple beam balance • Sanitized indirectly vented chemical-splash goggles • Chemical-resistant apron • Gloves • Lab 6 Reference Sheet

Safety Precautions

Even though there are food items used during this lab investigation, they should be treated as chemicals and not eaten. In addition, take the following safety precautions:

1. Notify the teacher immediately if you have an allergy to nuts.

LAB 6

2. Put on sanitized indirectly vented chemical-splash goggles and laboratory apron and gloves before starting the lab activity.

3. Never put consumables in your mouth.

4. Use caution when working with open flames while burning the nuts and using matches. Open flames can burn skin, and combustibles and flammables must be kept away from the open flame. If you have long hair, tie it back behind your head.

5. Handle all glassware with care to avoid breakage. Sharp glass can cut skin!

6. Wash your hands with soap and water after completing the lab activity.

Investigation Proposal Required?   ☐ Yes   ☐ No

Getting Started

During your investigation you will need to determine the amount of energy in each type of nut. To do this, you can use a calorimeter to measure the amount of heat energy released by burning each type of nut (see Figure L6.2). Measuring the temperature change in a sample of water will provide information related to the amount of energy stored in the nut.

To answer the guiding question, you will also need to consider other factors such as the cost of different types of nuts and how many calories it takes to complete various exercises. To accomplish this task, you must first determine what type of data you need to collect, how you will collect it, and how you will analyze it. To determine *what type of data you need to collect*, think about the following questions:

FIGURE L6.2 _____

A simple calorimeter that can be used to measure the amount of energy in a nut

- What information will tell you which nut(s) have the most energy?
- How will the calorimeter and thermometer (or temperature sensor) help you measure the energy in the nuts?
- What type of measurements or observations will you need to record during your investigation?

To determine *how you will collect your data*, think about the following questions:

- What will serve as a control (or comparison) condition?
- What types of treatment conditions will you need to set up and how will you do it?
- How often will you collect data and when will you do it?
- How will you make sure that your data are of high quality (i.e., how will you reduce error)?

- How will you keep track of the data you collect and how will you organize it?

To determine *how you will analyze your data*, think about the following questions:

- How will you determine if there is a difference between the treatment conditions and the control condition?
- What type of calculations will you need to make?
- What type of graph could you create to help make sense of your data?
- How many calories should your energy bar have?
- How much will your energy bar cost?

Connections to Crosscutting Concepts, the Nature of Science, and the Nature of Scientific Inquiry

As you work through your investigation, be sure to think about

- how energy and matter flow through living things while being totally conserved,
- how an object's shape or structure determines many of its properties and functions,
- the difference between data collected and evidence created in an investigation, and
- the role of imagination and creativity when solving problems in science.

Initial Argument

Once your group has finished collecting and analyzing your data, you will need to develop an initial argument. Your argument must include a claim, evidence to support your claim, and a justification of the evidence. The claim is your group's answer to the guiding question. The evidence is an analysis and interpretation of your data. Finally, the justification of the evidence is why your group thinks the evidence matters. The justification of the evidence is important because scientists can use different kinds of evidence to support their claims. Your group will create your initial argument on a whiteboard. Your whiteboard should include all the information shown in Figure L6.3.

FIGURE L6.3

Argument presentation on a whiteboard

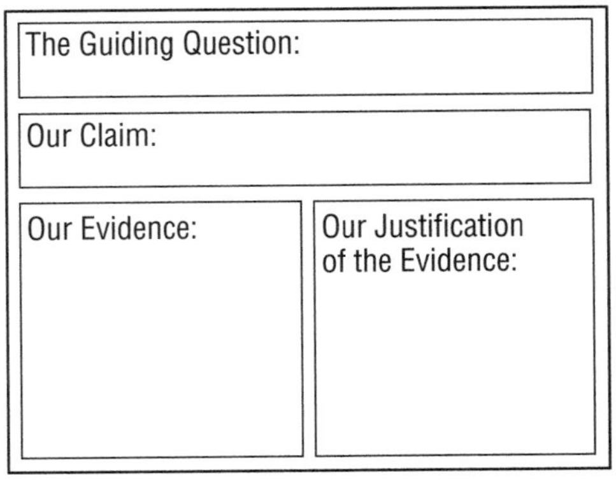

The Guiding Question:	
Our Claim:	
Our Evidence:	Our Justification of the Evidence:

Argumentation Session

The argumentation session allows all of the groups to share their arguments. One member of each group will stay at the lab station to share that group's argument, while the other members of the group go to the other lab stations one at a time to listen to and critique the arguments developed by their classmates. This is similar to how scientists present their arguments to other scientists at conferences. If you are responsible for critiquing your classmates' arguments, your goal is to look for mistakes so these mistakes can be fixed and they can make their argument better. The argumentation session is also a good time to think about ways you can make your initial argument better. Scientists must share and critique arguments like this to develop new ideas.

To critique an argument, you might need more information than what is included on the whiteboard. You will therefore need to ask the presenter lots of questions. Here are some good questions to ask:

- What did your group do to collect the data? Why do you think that way is the best way to do it?
- What did your group do to analyze the data? Why did your group decide to analyze it that way?
- What other ways of analyzing and interpreting the data did your group talk about?
- What did your group do to make sure that these calculations are correct?
- Why did your group decide to present your evidence in that way?
- What other claims did your group discuss before you decided on that one? Why did your group abandon those other ideas?
- How sure are you that your group's claim is accurate? What could you do to be more certain?

Once the argumentation session is complete, you will have a chance to meet with your group and revise your original argument. Your group might need to gather more data or design a way to test one or more alternative claims as part of this process. Remember, your goal at this stage of the investigation is to develop the most valid or acceptable answer to the research question!

Report

Once you have completed your research, you will need to prepare an investigation report that consists of three sections that provide answers to the following questions:

1. What question were you trying to answer and why?
2. What did you do during your investigation and why did you conduct your investigation in this way?
3. What is your argument?

Your report should answer these questions in two pages or less. This report must be typed, and any diagrams, figures, or tables should be embedded into the document. Be sure to write in a persuasive style; you are trying to convince others that your claim is acceptable or valid!

Lab 6 Reference Sheet

Costs and Exercise Calories

As you decide which nut will be best for your energy bar, be sure to consider how much of each type of nut would need to go into your bar and how much that will cost. The following table gives cost per pound for each type of nut used in this lab investigation.

Type of nut	Cost per pound
Peanut	$2.00
Cashew	$5.92
Pecan	$7.73
Almond	$8.88
Walnut	$10.43

Your energy bar should be made so that it stays together and maintains its shape. Many energy bars use honey or maple syrup as a binding agent to help the pieces stay together. The honey or syrup also adds flavor to the energy bar. It takes about 15 grams of honey or syrup per 100 grams of nuts for the energy bar to stay intact. The following table gives calories and cost for these binding agents.

Binding agent	Calories (per 100 g)	Cost (per 100 g)
Honey	304	$0.44
Maple syrup	260	$0.60

For your energy bar to be useful, it should supply enough calories that will provide energy to complete at least 30 minutes of exercise activity. The following table lists various activities and how much energy is used during 1 hour of that activity.

Activity	Calories (per hour)
Running (6 mph)	600
Walking (2 mph)	200
Biking (12 mph)	480
Swimming	420
Dancing	220
Aerobic exercise (high impact)	500
Martial arts	700
Jumping rope	850
Lifting weights	365

LAB 6

Lab 6. Energy in Food: Which Type of Nut Is Best for a New Energy Bar?

1. Steven made the claim that the energy we get from the food we eat can be traced all the way back to the Sun. Use what you know about energy transfer, photosynthesis, and cellular respiration to support his claim.

2. The energy we get from nutrients like carbohydrates and fats is a result of breaking chemical bonds and releasing the stored chemical potential energy. One gram of carbohydrate provides 4 food calories, but one gram of fat provides 9 food calories. Use what you know about energy and molecules to describe why fats provide more energy (per gram) than carbohydrates.

3. In the energy bar investigation, you collected data that you used to develop evidence.

   a. I agree with this statement.
   b. I disagree with this statement.

   Explain your answer, using an example from your investigation about energy in food.

4. Scientists do not need to be creative or have a good imagination to excel in science.

    a. I agree with this statement.
    b. I disagree with this statement.

    Explain your answer, using an example from your investigation about energy in food.

5. An important goal in science is to develop an understanding of how matter and energy move through complex systems. Explain why understanding the flow of matter and energy is important, using an example from your investigation about energy in food.

6. Scientists often investigate how the structure or composition of something is related to its function. Explain how composition and function are related, using an example from your investigation about energy in food.

LAB 7

Lab 7. Respiratory and Cardiovascular Systems: How Do Activity and Physical Factors Relate to Respiratory and Cardiovascular Fitness?

Purpose

The purpose of this lab is for students to *apply* their knowledge of the structure and function of body systems. Specifically, this investigation gives students an opportunity to investigate the relationship between the function of the cardiovascular and respiratory systems and students' activities and physical factors. Teachers will need to help students consider how to connect trends in these student activities to measurements of cardiovascular and respiratory rate. This lab gives students an opportunity to understand how living systems are studied through the development of system models and how the structure of living things influences their functions. Students will also have the opportunity to reflect on how science is influenced by society and culture and the specific role that experiments play in science.

The Content

The human body is a very complicated system. We take in a lot of materials, such as the air we breathe and the food and water we eat and drink. We also put out different materials, including gas, liquid, and solids. But these materials are very different from each other. Remember the law of conservation of mass, which tells us that matter cannot be created or destroyed; it can only change forms. So, obviously, the matter that enters our bodies goes through some interesting changes to result in the matter that comes out. In fact, the matter that comes in is used in many different ways to help our bodies grow and live.

All bodies, including those of humans, animals, plants, and bacteria, are made of cells. Some of these organisms (humans, animals, and plants) have their cells arranged into groups that all help the organisms to live. In humans and animals, scientists have identified several different *body systems*. Body systems are very organized groups of cells that serve specific functions that allow organisms to live. Although these systems are basically large groups of cells, those cells are organized into other levels. Pieces at each level work together to help pieces at the next level work. Body systems are made up of smaller units called *organs*, like the heart and lungs, which help the body take in and transport the nutrients it needs. Organs are made of up *tissues*, like muscle fibers in the heart, which help the organs function. For example, muscle tissue in the heart helps it to beat so it can transport materials through blood. Tissues are made up of *cells*, such as muscle cells that make up muscle fibers. Muscle cells shift proteins around to make muscle fibers grow and shrink so the whole muscle contracts causing the heart to beat, which helps the rest of the systems work.

The human body is made up of several systems. The functions these systems perform include breaking down food we eat (digestive), directing the activities of the body (nervous), and helping us fight off disease (immune). The *respiratory system* (Figure 7.1) includes the lungs and trachea that breathe in air, containing oxygen gas (O_2) for use in the body, and breathe out waste gases, like carbon dioxide gas (CO_2). The *cardiovascular system* (Figure 7.2; also called the circulatory system) uses blood, arteries, veins, and the heart to deliver materials such as O_2 from your lungs. It also removes waste, like CO_2, from cells; regulates body temperature; and helps the body fight off disease.

FIGURE 7.1 _____

Respiratory system

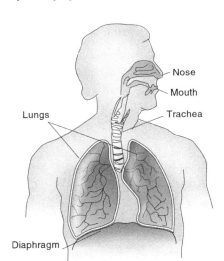

- Nose
- Mouth
- Lungs
- Trachea
- Diaphragm

FIGURE 7.2 _____

Cardiovascular system

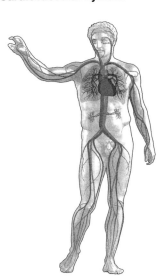

As you can see, the function of these two body systems is connected. They both work on moving gases into and out of the body. Cells in our body use O_2 to help make energy through a chemical process called *cellular respiration*. The same chemical process gives off CO_2, which is transported through blood from cells to the lungs to be breathed out. The gas exchange function of these two body systems helps all the cells in the body. This is because most cells must use O_2 to create energy.

During cellular respiration, O_2 helps convert the chemical energy in sugar molecules into a form animals can use. The energy from these molecules gets transferred by moving electrons from one molecule to another. When electrons are added or taken away, new chemical bonds and types of molecules can be formed. The O_2 helps transfer electrons from the chemical bonds in sugar to the chemical bonds in another molecule. The sugar molecules are broken up into smaller carbon chain molecules. As the carbon chains get rearranged during cellular respiration, a molecule of CO_2 is given off as a waste product.

Cellular respiration uses O_2 to release energy from molecules in food we eat. A waste product from this process is CO_2, the same gas that we breathe out. So as our body takes in and puts out these gases, so do the cells in our body. Indeed, that is why we breathe more when we exercise. We know that humans use respiration to produce energy because when a human breathes, the air that he or she inhales contains about 21% O_2 and less than 1% CO_2; however, when he or she exhales, the air contains about 15% O_2 and 5% CO_2. Our cells need more O_2 to make more energy to support the increased physical activity.

The functions of these systems are connected to the amount of work the body does. So personal activity levels, like being an athlete or doing regular exercise, can influence how well these systems work. Also, the structure of these systems is influenced by physical factors such as height or frame size. Larger people may have more cells than smaller people, and having more cells means higher energy demands. Physical activity can also increase the number of cells in the body. Gender (boy or girl) can influence the function of different systems. Environmental factors can also affect system functions; some locations have higher or lower amounts of O_2 available in the air. With less O_2 in the air, the body has to breathe more to get the O_2 it needs for activity.

Timeline

The instructional time needed to implement this lab investigation is 180–250 minutes. Appendix 2 (p. 355) provides options for implementing this lab investigation over several class periods. Option A (250 minutes) should be used if students are unfamiliar with scientific writing, because this option provides extra instructional time for scaffolding the writing process. You can scaffold the writing process by modeling, providing examples, and providing hints as students write each section of the report. Option B (180 minutes) should be used if students are familiar with scientific writing and have the skills needed to write an investigation report on their own. In option B, students complete stage 6 (writing the investigation report) and stage 8 (revising the investigation report) as homework.

Materials and Preparation

The materials needed to implement this investigation are listed in Table 7.1. You will need laptop computers to connect to and collect data from the sensors. These sensors can be ordered from a science supply company such as Vernier or Pasco. Vernier also sells a version of their heart rate monitor that is a handheld unit instead of a chest belt. Either version will work with the cardiovascular fitness protocol, which was adapted from the Vernier protocol supplied with these sensors.

TABLE 7.1

Materials list

Item	Quantity
Elastic straps (two sizes)	1 of each size per group
Transmitter belt	1 per group
Exercise heart rate monitor	1 per group
Timer	1 per group
Go!Link adaptor and laptop computer	1 per group
Stool, 1–2 feet high	1 per group
Investigation Proposal A (optional)	1 per group
Whiteboard, 2' × 3'*	1 per group
Lab Handout	1 per student
Lab 7 Reference Sheet: Cardiovascular Fitness Test Protocol and Tables	1 per group for protocol, 1 per student for table
Peer-review guide	1 per student
Checkout Questions	1 per student

* As an alternative, students can use computer and presentation software such as Microsoft PowerPoint or Apple Keynote to create their arguments.

Safety Precautions

Follow all normal lab safety rules. In addition, take the following safety precautions:

1. Before beginning this lab activity, check with the school nurse to make sure there are no potential student medical issues (heart or other health conditions) that may present a risk.

2. The heart rate monitor contains electrical connections covered by metal plates that are safe to touch skin. However, only use the heart rate monitor equipment as directed. Any playing or non-approved use of the equipment could potentially cause harm.

Topics for the Explicit and Reflective Discussion

Concepts That Can Be Used to Justify the Evidence

To provide an adequate justification of their evidence, students must explain why they included the evidence in their arguments and make the assumptions underlying their analysis and interpretation of the data explicit. In this investigation, students can use the following concepts to help justify their evidence:

- The cardiovascular and respiratory systems work together in the body to breathe in O_2 and breathe out CO_2.
- Cellular respiration uses O_2 to make energy for cells.
- The more activity a body does, the more energy its cells need.

We recommend that you review these concepts during the explicit and reflective discussion to help students make this connection.

How to Design Better Investigations

It is important for students to reflect on the strengths and weaknesses of the investigation they designed during the explicit and reflective discussion. Students should therefore be encouraged to discuss ways to eliminate potential flaws, measurement errors, or sources of bias in their investigations. To help students be more reflective about the design of their investigation, you can ask the following questions:

- What were some of the strengths of your investigation? What made it scientific?
- What were some of the weaknesses of your investigation? What made it less scientific?
- If you were to do this investigation again, what would you do to address the weaknesses in your investigation? What could you do to make it more scientific?

Crosscutting Concepts

This investigation is aligned with two crosscutting concepts found in *A Framework for K–12 Science Education,* and you should review these concepts during the explicit and reflective discussion.

- *Systems and system models:* It is critical for scientists to be able to define the system under study (e.g., the respiratory or cardiovascular system) and then make a model of it to understand it. Models can be physical, conceptual, or mathematical.
- *Structure and function:* In nature, the way a living thing is structured determines how it functions and places limits on what it can and cannot do.

The Nature of Science and the Nature of Scientific Inquiry

This investigation is aligned with two important concepts related to the *nature of science* (NOS) and the *nature of scientific inquiry* (NOSI), and you should review these concepts during the explicit and reflective discussion.

- *The influence of society and culture on science:* Science is influenced by the society and culture in which it is practiced because science is a human endeavor. Cultural values and expectations determine what scientists choose to investigate, how investigations are conducted, how research findings are interpreted, and what people see as implications. People also view some research as being more important than others because of cultural values and current events.

- *The nature and role of experiments:* Scientists use experiments to test the validity of a hypothesis (i.e., a tentative explanation) for an observed phenomenon. Experiments include a test and the formulation of predictions (expected results) if the test is conducted and the hypothesis is valid. The experiment is then carried out and the predictions are compared with the observed results of the experiment. If the predictions match the observed results, then the hypothesis is supported. If the predictions do not match the observed results, then the hypothesis is not supported. A signature feature of an experiment is the control of variables to help eliminate alternative explanations for observed results.

Hints for Implementing the Lab

- Have each group complete the cardiovascular fitness table for each student in that group, following the protocol. Groups can share their data after this is done.

- Let groups choose a physical or activity factor they want to include in their data analysis. Groups can help each other by collecting data about all the factors within their groups and sharing that information with other groups

- Use frame size as a factor instead of weight, because some students may be sensitive about their weight.

- Have groups compare cardiovascular rate with respiratory rate in light of the factor they are using.

Topic Connections

Table 7.2 (p. 122) provides an overview of the scientific practices, crosscutting concepts, disciplinary core ideas, and supporting ideas at the heart of this lab investigation. In addition, it lists NOS and NOSI concepts for the explicit and reflective discussion. Finally, it lists literacy and mathematics skills (*CCSS ELA* and *CCSS Mathematics*) that are addressed during the investigation.

LAB 7

TABLE 7.2

Lab 7 alignment with standards

Scientific practices	• Asking questions and defining problems • Developing and using models • Planning and carrying out investigations • Analyzing and interpreting data • Using mathematics and computational thinking • Constructing explanations • Engaging in argument from evidence • Obtaining, evaluating, and communicating information
Crosscutting concepts	• Systems and system models • Structure and function
Core idea	• LS1: From molecules to organisms: Structures and processes
Supporting ideas	• Cardiovascular system function • Respiratory system function • Cellular respiration • Energy needs for physical activity
NOS and NOSI concepts	• Social and cultural influences • Nature and role of experiments
Literacy connections (*CCSS ELA*)	• *Reading:* Key ideas and details, craft and structure, integration of knowledge and ideas • *Writing:* Text types and purposes, production and distribution of writing, research to build and present knowledge, range of writing • *Speaking and listening:* Comprehension and collaboration, presentation of knowledge and ideas
Mathematics connections (*CCSS Mathematics*)	• Make sense of problems and persevere in solving them • Reason abstractly and quantitatively • Construct viable arguments and critique the reasoning of others • Model with mathematics • Use appropriate tools strategically

Lab Handout

Lab 7. Respiratory and Cardiovascular Systems: How Do Activity and Physical Factors Relate to Respiratory and Cardiovascular Fitness?

Introduction

All bodies, including those of humans, animals, plants, and bacteria, are made of cells. Some of these organisms (humans, animals, and plants) have their cells arranged into groups that all help the organisms to live. In humans and animals, scientists have identified several different body systems. Body systems are very organized groups of cells that serve specific functions that allow organisms to live. Although these systems are basically large groups of cells, those cells are organized into other levels. Pieces at each level work together to help pieces at the next level work. Body systems are made up of smaller units called organs, like the heart and lungs, which help the body take in and transport the nutrients it needs. Organs are made of up tissues, like muscle fibers in the heart, which help the organs function. For example, muscle tissue in the heart helps it to beat so it can transport materials through blood. Tissues are made up of cells, such as muscle cells that make up muscle fibers. Muscle cells shift proteins around to make muscle fibers grow and shrink so the whole muscle contracts, causing the heart to beat, which helps the rest of the systems work.

The human body is made up of several systems. The functions these systems perform include breaking down food we eat (digestive), directing the activities of the body (nervous), and helping us fight off disease (immune). The respiratory system (Figure L7.1) includes the lungs and trachea that breathe in air, containing oxygen (O_2) for use in the body, and breathe out waste gases, like carbon dioxide (CO_2). The cardiovascular system (Figure L7.2; also called the circulatory system) uses blood, arteries, veins, and

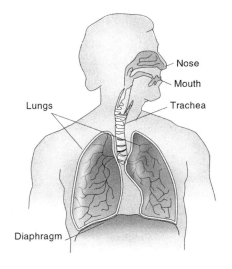

FIGURE L7.1 _____

Respiratory system

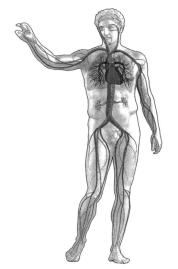

FIGURE L7.2 _____

Cardiovascular system

the heart to deliver materials such as O_2 from your lungs. It also removes waste, like CO_2, from cells; regulates body temperature; and helps the body fight off disease.

As you can see, the function of these two body systems is connected. They both work on moving gases into and out of the body. Cells in our body use O_2 to help make energy through a chemical process called cellular respiration. The same chemical process gives off CO_2, which is transported through blood from cells to the lungs to be breathed out. The gas exchange function of these two body systems helps all the cells in the body. This is because most cells must use O_2 to create energy.

Cellular respiration uses O_2 to release energy from molecules in food we eat. A waste product from this process is CO_2, the same gas that we breathe out. So as our body takes in and puts out these gases, so do the cells in our body. Indeed, that is why we breathe more when we exercise. We know that humans use respiration to produce energy because when a human breathes, the air that he or she inhales contains about 21% O_2 and less than 1% CO_2; however, when he or she exhales, the air contains about 15% O_2 and 5% CO_2. Our cells need more O_2 to make more energy to support the increased physical activity.

The functions of these systems are connected to the amount of work the body does. So personal activity levels, like being an athlete or doing regular exercise, can influence how well these systems work. Also, the structure of these systems is influenced by physical factors such as height or frame size. Larger people may have more cells than smaller people, and having more cells means higher energy demands. Physical activity can also increase the number of cells in the body. Gender (boy or girl) can influence the function of different systems. Environmental factors can also affect the functions of the body; some locations have higher or lower amounts of O_2 available in the air. With less O_2 in the air, the body has to breathe more to get the O_2 it needs for activity. The function of these body systems is tied to the energy needs of the body.

Your Task

Design an investigation to explore the relationship between fellow students' cardiovascular and respiratory fitness and different activity and physical factors.

The guiding question of this investigation is, **How do activity and physical factors relate to respiratory and cardiovascular fitness?**

Materials

You may use any of the following materials during your investigation:

- Elastic straps
- Transmitter belt
- Exercise heart rate monitor
- Timer
- Go!Link adaptor and laptop computer

- Stool (1–2 feet high)
- Cardiovascular fitness protocol
- Cardiovascular fitness table
- Lab 7 Reference Sheet (1 per student)

Safety Precautions

Follow all normal lab safety rules. In addition, take the following safety precautions:

1. Notify your teacher if you have heart or other health conditions.

2. The heart rate monitor contains electrical connections covered by metal plates that are safe to touch your skin. However, only use the heart rate monitor equipment as directed. Any playing or non-approved use of the equipment could potentially hurt you or your classmates.

Investigation Proposal Required?   ☐ Yes   ☐ No

Getting Started

You will be given a Lab 7 Reference Sheet that includes a cardiovascular fitness protocol and a cardiovascular fitness table; these resources will guide the data collection process. After reviewing this sheet, you will need to determine what type of data you need to collect, how you will collect the data, and how you will analyze the data. To determine *what type of data you need to collect*, think about the following questions:

- What personal or physical factors could influence respiratory and cardiovascular fitness?
- What type of measurements or observations will you need to record during your investigation?

To determine *how you will collect your data*, think about the following questions:

- What types of treatment conditions will you need to set up and how will you do it?
- During the experiment, when will you collect data and how often will you collect it?
- How will you make sure that your data are of high quality (i.e., how will you reduce error)?
- How will you keep track of the data you collect and how will you organize it?

To determine *how you will analyze your data*, think about the following questions:

- How will you determine overall cardiovascular fitness?
- How will you compare subgroups?
- What type of calculations will you need to make?
- What type of graph could you create to help make sense of your data?

LAB 7

Connections to Crosscutting Concepts, the Nature of Science, and the Nature of Scientific Inquiry

As you work through your investigation, be sure to think about

- how scientists study systems by creating models of their structure and function,
- how the structure of living things affects the way they function,
- the way science is influenced by the society in which it takes place, and
- the specific role that experiments play in science.

Initial Argument

Once your group has finished collecting and analyzing your data, you will need to develop an initial argument. Your argument must include a claim, evidence to support your claim, and a justification of the evidence. The claim is your group's answer to the guiding question. The evidence is an analysis and interpretation of your data. Finally, the justification of the evidence is why your group thinks the evidence matters. The justification of the evidence is important because scientists can use different kinds of evidence to support their claims. Your group will create your initial argument on a whiteboard. Your whiteboard should include all the information shown in Figure L7.3.

FIGURE L7.3

Argument presentation on a whiteboard

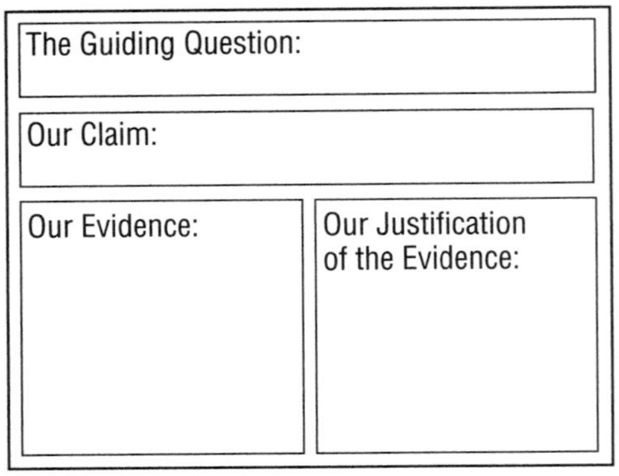

The Guiding Question:	
Our Claim:	
Our Evidence:	Our Justification of the Evidence:

Argumentation Session

The argumentation session allows all of the groups to share their arguments. One member of each group will stay at the lab station to share that group's argument, while the other members of the group go to the other lab stations one at a time to listen to and critique the arguments developed by their classmates. This is similar to how scientists present their arguments to other scientists at conferences. If you are responsible for critiquing your classmates' arguments, your goal is to look for mistakes so these mistakes can be fixed and they can make their argument better. The argumentation session is also a good time to think about ways you can make your initial argument better. Scientists must share and critique arguments like this to develop new ideas.

To critique an argument, you might need more information than what is included on the whiteboard. You will therefore need to ask the presenter lots of questions. Here are some good questions to ask:

- What did your group do to collect the data? Why do you think that way is the best way to do it?

- What did your group do to analyze the data? Why did your group decide to analyze it that way?

- What other ways of analyzing and interpreting the data did your group talk about?

- What did your group do to make sure that these calculations are correct?

- Why did your group decide to present your evidence in that way?

- What other claims did your group discuss before you decided on that one? Why did your group abandon those other ideas?

- How sure are you that your group's claim is accurate? What could you do to be more certain?

Once the argumentation session is complete, you will have a chance to meet with your group and revise your original argument. Your group might need to gather more data or design a way to test one or more alternative claims as part of this process. Remember, your goal at this stage of the investigation is to develop the most valid or acceptable answer to the research question!

Report

Once you have completed your research, you will need to prepare an investigation report that consists of three sections that provide answers to the following questions:

1. What question were you trying to answer and why?

2. What did you do during your investigation and why did you conduct your investigation in this way?

3. What is your argument?

Your report should answer these questions in two pages or less. This report must be typed, and any diagrams, figures, or tables should be embedded into the document. Be sure to write in a persuasive style; you are trying to convince others that your claim is acceptable or valid!

LAB 7

Lab 7 Reference Sheet

Cardiovascular Fitness Test Protocol and Tables

Cardiovascular Fitness Test Protocol

To determine the cardiovascular health of a person, you will need to measure reclining heart rate, exercise heart rate (step test), and exercise recovery time.[1]

Reclining Heart Rate

1. Have the subject lie on a clean surface or table. Begin collecting heart rate data. While the subject is lying on the table, count the number of breaths he or she takes in one minute.

2. Record the subject's heart rate after two minutes in the cardiovascular fitness table.

3. Assign fitness points to the subject based on Table R7.1 and record the value in the cardiovascular fitness table. Also, record the subject's respiratory rate in the cardiovascular fitness table.

TABLE R7.1 _____

Reclining heart rate fitness points

Beats/min.	Fitness points
50–60	6
61–70	5
71–80	4
81–90	3
91–100	2
101–110	1
> 110	0

---

1 This protocol has been adapted from a similar protocol available from Vernier, Inc., for use with their heart rate monitors.

Step Test

4. Before performing the step test, have the subject stand still for 30 seconds. Record the subject's heart rate at 30 seconds as the subject's pre-exercise heart rate.

5. Perform a step test using the following procedure:

   a. Place the right foot on the top step of the stool.

   b. Place the left foot completely on the top step of the stool next to the right foot.

   c. Place the right foot back on the floor.

   d. Place the left foot completely on the floor next to the right foot.

   e. Repeat the cycle as quickly as possible for 30 seconds.

6. Record the heart rate in the subject's cardiovascular fitness table. **DO NOT STOP DATA COLLECTION!** Start timer and quickly move to step 7. While working on steps 7 and 8 for heart rate, measure the subject's *respiratory* rate in 30 seconds and multiply that number by 2. (You can start step 7 and then begin measuring respiratory rate.)

Exercise Recovery Time

7. Have the subject remain standing and keep relatively still. Monitor the heart rate readings and stop timing when the readings return to the pre-exercise heart rate value recorded in step 4. Record the recovery time in the cardiovascular fitness table.

8. Locate the subject's recovery time in Table R7.2 and record the corresponding fitness point value in the cardiovascular fitness table. Also, record the subject's respiratory rate when he or she has reached recovery heart rate.

TABLE R7.2

Recovery time fitness points

Time (sec.)	Fitness points
0–30	6
31–60	5
61–90	4
91–120	3
121–150	2
151–180	1
>180	0

LAB 7

9. Subtract the subject's pre-exercise heart rate from his or her heart rate after five stepping cycles of exercise. Record this heart rate increase in the endurance row of the cardiovascular fitness table. Do the same subtraction for the respiratory rate.

10. Locate the row corresponding to the pre-exercise heart rate in Table R7.3 and use the heart rate increase value to determine endurance fitness points. Record the subject's endurance fitness points in the cardiovascular fitness table.

TABLE R7.3

Endurance fitness points

Reclining heart rate (beats/min.)	Heart rate increase after exercise (beats/min.)				
	0–10	11–20	21–30	31–40	41+
50–60	6	5	4	3	2
61–70	5	4	3	2	1
71–80	4	3	2	1	0
81–90	3	2	1	0	0
91–100	2	1	0	0	0
101–110	1	0	0	0	0
>110	0	0	0	0	0

Cardiovascular Fitness Table

Subject number: _____   Gender: _____   Age: _____

Frame size: _____   Height: _____   Factor: _____

Condition	Value	Fitness points	Respiratory rate
Reclining heart rate	beats/min.		breaths/min.
Pre-exercise heart rate	beats/min.		breaths/min.
Step test	beats/min.		breaths/min.
Exercise recovery time	seconds		
Endurance	beats/min.		breaths/min.
		Total:	

LAB 7

Lab 7. Respiratory and Cardiovascular Systems: How Do Activity and Physical Factors Relate to Respiratory and Cardiovascular Fitness?

There are many systems in the body that must work together to sustain life. Systems such as the respiratory system, circulatory system, and digestive system are involved in making sure our bodies have the energy it needs to function.

1. Use what you know about energy transfer and each of these systems to describe how they work together to help our body function.

2. Jeremy made the claim in science class that most body systems work together, but the nervous system is the only one that operates in isolation. Do you agree with Jeremy? Explain your reasoning.

3. Scientists should not allow their society or culture to influence their work.

   a. I agree with this statement.

   b. I disagree with this statement.

   Explain your answer, using an example from your investigation about the respiratory and cardiovascular systems.

4. Investigations in medical science often involve people, so experiments are not used.

    a. I agree with this statement.

    b. I disagree with this statement.

    Explain your answer, using an example from your investigation about the respiratory and cardiovascular systems.

5. Scientists often generate models when they are working with complex systems or events. Explain why using models in science is helpful, using an example from your investigation about the respiratory and cardiovascular systems.

6. Scientists often investigate how the structure of something is related to its function. Explain how structure and function are related, using an example from your investigation about the respiratory and cardiovascular systems.

LAB 8

Lab 8. Memory and Stimuli: How Does the Way Information Is Presented Affect Working Memory?

Purpose

The purpose of this lab is for students to *apply* their knowledge of the structure and function of body systems. Specifically, this investigation gives students an opportunity to explore the relationship between patterns in letters and numbers and how much people can remember. Teachers will need to help students consider how patterns can be found in sequences of numbers and letters. This lab gives students an opportunity to understand how scientists focus on finding patterns across structures and events in the natural world and how the structure of living things influences their functions. Students will also have the opportunity to reflect on how science is influenced by society and culture and how scientists use imagination and creativity in solving problems.

The Content

The human body is made up of several systems that work together. These systems perform certain activities necessary for living. The respiratory system, including the lungs, allows us to take in oxygen and get rid of carbon dioxide. The circulatory system uses the heart, blood, and blood vessels to move chemicals around the body. The immune system uses special cells and chemicals to fight off germs and disease. The digestive system, including the stomach, allows us to break down and use the food we eat. The *nervous system* is responsible for taking in information and directing the actions of other parts of the body. This system includes nerves and the brain. The nerves carry information around the body and the brain. The brain sends and receives information, controlling many different activities at the same time.

The brain gets information about the world around the body through the *senses*. There are five senses: hearing, seeing, tasting, smelling, and touching. Each of these senses uses a special organ that is filled with a large number of nerves. The large number of nerves around these organs allows them to take in lots of specific information. Those nerves send information to the brain that tells us what is going on in the world. Hearing involves the ears, seeing involves the eyes, tasting involves the tongue, smelling involves the nose, and touching involves the skin.

The eyes take in light waves to send information to the brain. The brain uses that information to understand the world around us. As that information is processed, the brain makes connections across the information and organizes it based on patterns. Using those patterns, the brain is able to take in more information faster and retain it longer. Thus, the presence of recognizable patterns, whether in structure or activity, makes it easier for our brains to make sense of the world.

Our senses respond to *stimuli* (the singular form is *stimulus*), which are things or events that evoke a specific functional reaction in an organ or tissue. All things our eyes see, including the words on this page, are stimuli. Our senses take in information about the stimuli and send it to the brain by chemical and electrical signals. Our brain and nerves create pathways that are used each time that information is perceived by our senses. Our brain reads that information and acts on it. The brain reads and uses the information in several ways:

- *Short-term memory* keeps information for only a few seconds. Your short-term memory handles the most basic information, like the light level in a room.

- *Working memory* keeps information for just a little longer, allowing us to organize it and make sense of it. Your working memory is being used as you read this sentence.

- *Long-term memory* keeps a large amount of information for long periods of times, up to years and decades.

All of these types of memories work together as the brain processes information taken in and responds back to it through various actions. The brain takes in and responds almost instantaneously to massive amounts of information. Some information becomes so routine that memory beyond short term is not activated.

Timeline

The instructional time needed to implement this lab investigation is 130–200 minutes. Appendix 2 (p. 355) provides options for implementing this lab investigation over several class periods. Option C (200 minutes) should be used if students are unfamiliar with scientific writing, because this option provides extra instructional time for scaffolding the writing process. You can scaffold the writing process by modeling, providing examples, and providing hints as students write each section of the report. Option D (130 minutes) should be used if students are familiar with scientific writing and have the skills needed to write an investigation report on their own. In option D, students complete stage 6 (writing the investigation report) and stage 8 (revising the investigation report) as homework.

Materials and Preparation

The materials needed to implement this investigation are listed in Table 8.1 (p. 136). The memory letter cards are available *www.nsta.org/publications/press/extras/adi-lifescience.aspx*. The number cards can be made or you can use a deck of playing cards.

LAB 8

TABLE 8.1

Materials list

Item	Quantity
Set of number cards (1–9)	1 per group
Set of memory letter cards	3 per group
Paper	4 sheets per group
Timer	1 per group
Investigation Proposal C (optional)	1 per group
Whiteboard, 2' × 3'*	1 per group
Lab Handout	1 per student
Peer-review guide	1 per student
Checkout Questions	1 per student

* As an alternative, students can use computer and presentation software such as Microsoft PowerPoint or Apple Keynote to create their arguments.

Safety Precautions

Follow all normal lab safety rules.

Topics for the Explicit and Reflective Discussion

Concepts That Can Be Used to Justify the Evidence

To provide an adequate justification of their evidence, students must explain why they included the evidence in their arguments and make the assumptions underlying their analysis and interpretation of the data explicit. In this investigation, students can use the following concepts to help justify their evidence:

- The brain receives information through our senses.
- Our senses rely on special organs that take in certain types of information.
- Different orders of letters or numbers are different structures of visual stimuli.
- Patterns help the brain process information faster.
- Three types of memory exist for the human brain.

We recommend that you review these concepts during the explicit and reflective discussion to help students make this connection.

How to Design Better Investigations

It is important for students to reflect on the strengths and weaknesses of the investigation they designed during the explicit and reflective discussion. Students should therefore be encouraged to discuss ways to eliminate potential flaws, measurement errors, or sources of bias in their investigations. To help students be more reflective about the design of their investigation, you can ask the following questions:

- What were some of the strengths of your investigation? What made it scientific?

- What were some of the weaknesses of your investigation? What made it less scientific?

- If you were to do this investigation again, what would you do to address the weaknesses in your investigation? What could you do to make it more scientific?

Crosscutting Concepts

This investigation is aligned with two crosscutting concepts found in *A Framework for K–12 Science Education,* and you should review these concepts during the explicit and reflective discussion.

- *Patterns:* Observed patterns in nature (e.g., all living things are composed of at least one cell) guide organization and classification systems in biology.

- *Structure and function:* In nature, the way a living thing is structured determines how it functions and places limits on what it can and cannot do.

The Nature of Science and the Nature of Scientific Inquiry

This investigation is aligned with two important concepts related to the *nature of science* (NOS) and the *nature of scientific inquiry* (NOSI), and you should review these concepts during the explicit and reflective discussion.

- *The influence of society and culture on science:* Science is influenced by the society and culture in which it is practiced because science is a human endeavor. Cultural values and expectations determine what scientists choose to investigate, how investigations are conducted, how research findings are interpreted, and what people see as implications. People also view some research as being more important than others because of cultural values and current events.

- *The importance of imagination and creativity in science:* Students should learn that developing explanations for or models of natural phenomena and then figuring out how they can be put to the test of reality is as creative as writing poetry, composing music, or designing skyscrapers. Scientists must also use their imagination and creativity to figure out new ways to test ideas and collect or analyze data.

LAB 8

Hints for Implementing the Lab

- Students being tested should not use the same set of cards they used when testing another student.
- You can make more letter cards if you think they are needed. The pattern on Card 1 is with five three-letter words in sequence. The pattern on Card 2 uses the same letters to create the following sequence: two random letters + new three-letter word + five random letters + new three-letter word + two random letters. The letters on Card 3 are just randomly ordered, but try to avoid indirectly making words in this sequence.
- Have groups share memory data to create larger data sets for analysis.

Topic Connections

Table 8.2 provides an overview of the scientific practices, crosscutting concepts, disciplinary core ideas, and supporting ideas at the heart of this lab investigation. In addition, it lists NOS and NOSI concepts for the explicit and reflective discussion. Finally, it lists literacy and mathematics skills (*CCSS ELA* and *CCSS Mathematics*) that are addressed during the investigation.

TABLE 8.2

Lab 8 alignment with standards

Scientific practices	• Asking questions and defining problems • Planning and carrying out investigations • Analyzing and interpreting data • Constructing explanations • Engaging in argument from evidence • Obtaining, evaluating, and communicating information
Crosscutting concepts	• Patterns • Structure and function
Core idea	• LS1: From molecules to organisms: Structures and processes
Supporting ideas	• Senses and their organs • Stimuli structure • Types of memory • Brain
NOS and NOSI concepts	• Social and cultural influences • Imagination and creativity in science
Literacy connections (*CCSS ELA*)	• *Reading:* Key ideas and details, craft and structure, integration of knowledge and ideas • *Writing:* Text types and purposes, production and distribution of writing, research to build and present knowledge, range of writing • *Speaking and listening:* Comprehension and collaboration, presentation of knowledge and ideas
Mathematics connections (*CCSS Mathematics*)	• Make sense of problems and persevere in solving them • Reason abstractly and quantitatively • Construct viable arguments and critique the reasoning of others • Look for and express regularity in repeated reasoning

Lab Handout

Lab 8. Memory and Stimuli: How Does the Way Information Is Presented Affect Working Memory?

Introduction

The human body is made up of several systems that work together. These systems perform certain activities necessary for living. For example, the respiratory system (which includes the lungs) allows us to take in oxygen and get rid of carbon dioxide, and the digestive system (which includes the stomach) allows us to break down and use the food we eat. The *nervous system* is responsible for taking in information and directing the actions of other parts of the body. This system includes nerves and the brain. The nerves carry information around the body and the brain. The brain sends and receives information, controlling many different activities at the same time.

The brain gets information about the world around the body through *senses*. There are five senses: hearing, seeing, tasting, smelling, and touching. Each of these senses uses a special organ that is filled with lots of nerves. Those nerves send information to the brain that tells us what is going on in the world. The *eyes* are the organ for seeing. The eyes take in light waves to send information to the brain, and the brain uses that information to understand the world around us. But what does the brain do with that information?

Our senses respond to *stimuli*, which is the plural of *stimulus*. A stimulus is a thing or event that evokes a specific functional reaction in an organ or tissue. All things our eyes see, including the words on this page, are stimuli. Our senses take in information about the stimuli and send it to the brain by chemical and electrical signals. The brain reads that information and acts on it. The brain reads and uses the information in several ways:

- *Short-term memory* keeps information for only a few seconds. Your short-term memory handles the most basic information, like the light level in a room.
- *Working memory* keeps information for just a little longer, allowing us to organize it and make sense of it. You are using your working memory as you read this sentence.
- *Long-term memory* keeps lots of information for long periods of times, up to years and decades.

As information is processed, the brain makes connections across it and organizes it based on patterns. Using those patterns, the brain is able to take in more information faster and retain it longer. Information that is presented in a pattern makes it easier for our brains to make sense of the world.

Your Task

Design an investigation to see how much information people can store in their working memory. Your goal is to explore how the amount of information and the order in which it is presented affects what people can remember.

The guiding question of this investigation is, **How does the way information is presented affect working memory?**

Materials

You may use any of the following materials during your investigation:

- Set of cards numbered 1 through 9
- Set of memory letter cards
- Paper
- Timer

Safety Precautions

Follow all normal lab safety rules.

Investigation Proposal Required?   ☐ Yes   ☐ No

Getting Started

You will be given a set of cards to see how many numbers people can remember. For this test, you lay out one numbered card and give a person 20 seconds to memorize it. Cover the card with a piece of paper, and then have the person tell you the number. Lay out another card beside the first one. Give the person another 20 seconds to memorize the two numbers. Cover the card and then have the person tell you the numbers in the correct order. Keep adding more cards until the person cannot tell you the numbers in the correct order.

You will also be given a set of cards with the same group of letters on them but in different orders. For this test, lay out Card 1 in the set and give a person 30 seconds to memorize it. Take the card away and have the person tell you the letters that were on the card. Lay out Card 2 in the set and give a person 30 seconds to memorize it. Take the card away and have the person tell you the letters that were on the card. Do the same thing using Card 3.

To answer the guiding question, you will need to determine what type of data you need to collect, how you will collect the data, and how you will analyze the data. To determine *what type of data you need to collect*, think about the following questions:

- What kind of information can you get from the person telling you about what's on the card?

LAB 8

- What type of measurements or observations will you need to record during your investigation?

To determine *how you will collect the data*, think about the following questions:

- What types of conditions will you need to set up and how will you do it?
- During the experiment, when will you collect data and how often will you collect it?
- How will you make sure that your data are of high quality (i.e., how will you reduce error)?
- How will you keep track of the data you collect and how will you organize it?

To determine *how you will analyze the data*, think about the following questions:

- How will you connect information about people's memories to the types of cards you used?
- How will you compare subgroups?
- What type of calculations will you need to make?
- What type of graph could you create to help make sense of your data?

Connections to Crosscutting Concepts, the Nature of Science, and the Nature of Scientific Inquiry

As you work through your investigation, be sure to think about

- how scientists look for patterns in the world,
- how the structure of living things affects the way they function,
- the way science is influenced by the society in which it takes place, and
- the role of imagination and creativity when solving problems in science.

Initial Argument

Once your group has finished collecting and analyzing your data, you will need to develop an initial argument. Your argument must include a claim, evidence to support your claim, and a justification of the evidence. The claim is your group's answer to the guiding question. The evidence is an analysis and interpretation of your data. Finally, the justification of the evidence is why your group thinks the evidence matters. The justification of the evidence is important because scientists can use different kinds of evidence to support their claims. Your group will create your initial argument on a whiteboard. Your whiteboard should include all the information shown in Figure L8.1.

Argumentation Session

The argumentation session allows all of the groups to share their arguments. One member of each group will stay at the lab station to share that group's argument, while the other members of the group go to the other lab stations one at a time to listen to and critique the arguments developed by their classmates. This is similar to how scientists present their arguments to other scientists at conferences. If you are responsible for critiquing your classmates' arguments, your goal is to look for mistakes so these mistakes can be fixed and they can make their argument better. The argumentation session is also a good time to think about ways you can make your initial argument better. Scientists must share and critique arguments like this to develop new ideas.

FIGURE L8.1

Argument presentation on a whiteboard

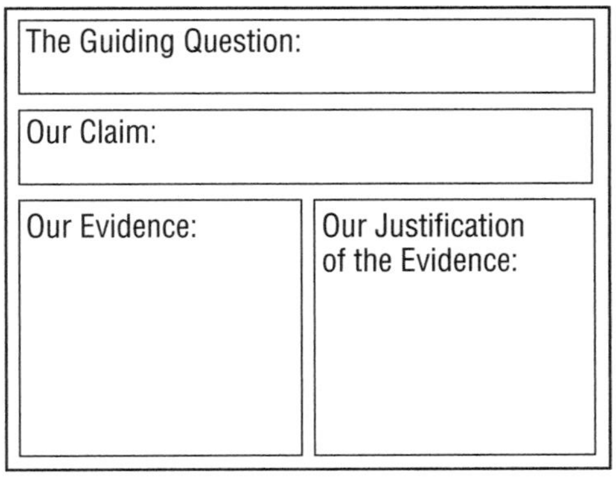

To critique an argument, you might need more information than what is included on the whiteboard. You will therefore need to ask the presenter lots of questions. Here are some good questions to ask:

- What did your group do to collect the data? Why do you think that way is the best way to do it?
- What did your group do to analyze the data? Why did your group decide to analyze it that way?
- What other ways of analyzing and interpreting the data did your group talk about?
- What did your group do to make sure that these calculations are correct?
- Why did your group decide to present your evidence in that way?
- What other claims did your group discuss before you decided on that one? Why did your group abandon those other ideas?
- How sure are you that your group's claim is accurate? What could you do to be more certain?

Once the argumentation session is complete, you will have a chance to meet with your group and revise your original argument. Your group might need to gather more data or design a way to test one or more alternative claims as part of this process. Remember, your goal at this stage of the investigation is to develop the most valid or acceptable answer to the research question!

Report

Once you have completed your research, you will need to prepare an investigation report that consists of three sections that provide answers to the following questions:

LAB 8

1. What question were you trying to answer and why?

2. What did you do during your investigation and why did you conduct your investigation in this way?

3. What is your argument?

Your report should answer these questions in two pages or less. This report must be typed, and any diagrams, figures, or tables should be embedded into the document. Be sure to write in a persuasive style; you are trying to convince others that your claim is acceptable or valid!

Checkout Questions

Lab 8. Memory and Stimuli: How Does the Way Information Is Presented Affect Working Memory?

1. Phone numbers are organized in a 3-3-4 pattern, for example: 555-867-5309. Use what you know about working memory to explain why organizing phone numbers in this way is more helpful that listing all 10 numbers in a row.

2. The two images below were made with six small sticks. If a person has only has one second to look at the images and then has to redraw them, which image do you think would be easier to remember? Explain your reasoning.

Image A

Image B

3. Science is influenced by the society in which it takes place.

   a. I agree with this statement.

   b. I disagree with this statement.

Explain your answer, using an example from your investigation about memory and stimuli.

4. Imagination and creativity have no place in scientific investigations.

   a. I agree with this statement.
   b. I disagree with this statement.

   Explain your answer, using an example from your investigation about memory and stimuli.

5. Scientists often look for patterns when they are investigating the world. Explain why identifying patterns in the world is helpful, using an example from your investigation about memory and stimuli.

6. Scientists often investigate how the structure of something is related to its function. Explain how structure and function are related in terms of body systems, using an example from your investigation about memory and stimuli.

SECTION 3

Life Sciences
Core Idea 2

Ecosystems: Interactions, Energy, and Dynamics

Introduction Labs

LAB 9

Teacher Notes

Lab 9. Population Growth: What Factors Limit the Size of a Population of Yeast?

Purpose

The purpose of this lab is to *introduce* students to the concept of population growth and some of the factors that can affect the size of a population. Specifically, this investigation gives students an opportunity to explore how the size of a population of yeast changes in an experiment that involves control conditions and the manipulation of one variable. This lab gives students an opportunity to understand how scientists look for patterns in the natural world and how living things undergo periods of stability and change. Students will also have the opportunity to reflect on the difference between observations and inferences and the role of experiments in science.

The Content

A *population* is a group of individuals that belong to the same species and live in the same region at the same time. In 1800, the total number of people on Earth was about 1 billion. Now the world population is 7 billion. This observation has caused many scientists to ask questions such as "What caused this dramatic increase in the human population?" and " How long will the human population continue to grow?" Human population growth has been a "hot topic" of discussion in the popular media and among ecologists (scientists who study how organisms interact with each other and the environment). These questions are good ways to begin discussing the ideas addressed during this lab investigation and help students make a personal connection to the material. Other relevant topics that can help introduce the ideas about population dynamics include the need for hunting and fishing seasons and the influence of urbanization on animal populations. Figure 9.1 provides a detailed graphic that highlights some of the major implications for a rapidly growing human population.

To better understand human population growth on Earth, it is imperative to study *population dynamics*, or the changing size, density, and range of a population. *Population density* refers to the concentration of individuals in a specific population in a specified area, and *population range* refers to the size and spread of the specified area that the population inhabits. These different measures of populations are important for understanding population dynamics. Scientists studying population dynamics investigate how different factors in living organisms and the environment shape those changes.

Humans are not the only living things studied through population dynamics. This distinction is important to emphasize to students so they do not develop the misconception

FIGURE 9.1

Web of potential effects due to rapid human population growth

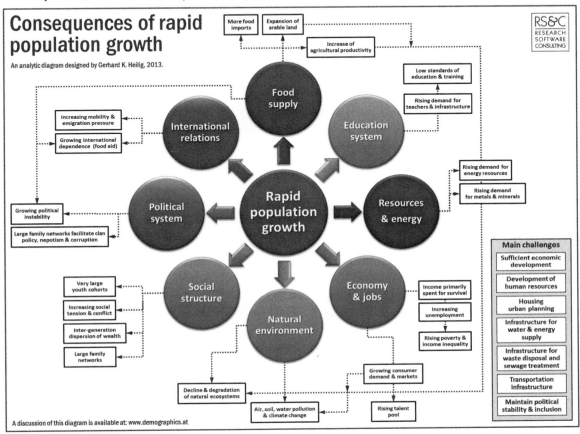

that humans are the only organisms organized in populations. Furthermore, students should understand that patterns found in the population dynamics of other species reflect similar patterns that occur in human populations.

Again, population dynamics involves living organisms and the environment in which they live. Factors related to all living organisms can include what they eat and drink, the amount of space they need, and how they interact with other organisms. Many of these factors are considered *biotic factors*, because they involve living things in an ecosystem. Factors related to the environment can include how much space is available and the weather and climate patterns in a particular area. Many of these factors are considered *abiotic factors*, because they involve nonliving pieces of an ecosystem. Some care should be taken when discussing the examples for this concept. Biotic factors must involve something living, so while the space taken up by a population of organisms involves living things, that space is not a living thing itself and incorporates abiotic elements as well.

By investigating these biotic and abiotic factors, scientists determine important relationships that help populations grow or reduce the number of organisms in an area. Scientists study the population dynamics of many different organisms to help us understand more about what affects human populations. Models and model organisms are extensively used in biological research to explore these relationships, which is another crosscutting concept that can be referenced when discussing this activity. Scientists will often investigate the effects of certain interventions on other animal populations before attempting similar studies in humans. Also, scientists will study how human populations change other populations of organisms. Studies of this kind are very important to helping society understand sustainability issues, such as overfishing and deforestation, and larger trends such as climate change. Knowing about these relationships helps scientists and policy makers make decisions about how to manage natural resources and organisms in many different types of ecosystems.

In conducting this lab investigation, all groups should see some growth in their yeast population, regardless of the test conditions used. Because of the variety of factors tested, there will not be a single controlling factor that affects growth. Populations of yeast may grow in response to treatment conditions for the first day or so, but then there may be a leveling off of the growth because of the limiting factors. Once the resources are used up, the population will no longer grow because it has reached its *carrying capacity*. Carrying capacity refers to the population size of a species that a certain environment with certain levels of resources can sustain indefinitely. Although this concept is not mentioned in the Lab Handout, teachers use the explicit and reflective discussion to introduce this vocabulary and definition.

Timeline

The instructional time needed to implement this lab investigation is 180–250 minutes. Appendix 2 (p. 355) provides options for implementing this lab investigation over several class periods. Option E (250 minutes) should be used if students are unfamiliar with scientific writing, because this option provides extra instructional time for scaffolding the writing process. You can scaffold the writing process by modeling, providing examples, and providing hints as students write each section of the report. Option F (180 minutes) should be used if students are familiar with scientific writing and have the skills needed to write an investigation report on their own. In option F, students complete stage 6 (writing the investigation report) and stage 8 (revising the investigation report) as homework.

Note: The timeline options in Appendix 2 refer to total class days needed for the investigation. However, due to the nature of the data that will be collected in this lab activity (minimum of three days of microscope cell counts), it is recommended that you allow individual students from each group to collect the data while the rest of the class engages in other activities. Students could also be asked to perform data collection outside of the class period.

Materials and Preparation

The materials needed to implement this investigation are listed in Table 9.1. The yeast cultures can be purchased at most grocery stores through commercial brands. Follow the directions on the product package to activate the yeast.

TABLE 9.1

Materials list

Item	Quantity
Consumables	
Yeast culture	5 ml per group
Sugar solution	15 ml per group
Iodine solution	1 ml per group
Distilled water	20 ml per group
Equipment	
Graduated pipette, 1 ml	5 per group
Test tubes	3–4 per group
Test tube rack	1 per group
Simple compound light microscope	1 per group
Microscope slides	15 per group
Cover slips	15 per group
Sanitized indirectly vented chemical-splash goggles	1 per student
Chemical-resistant apron	1 per student
Gloves	1 pair per student
Investigation Proposal A (optional)	1 per group
Whiteboard, 2' × 3'*	1 per group
Lab Handout	1 per student
Peer-review guide	1 per student
Checkout Questions	1 per student

* As an alternative, students can use computer and presentation software such as Microsoft PowerPoint or Apple Keynote to create their arguments.

LAB 9

Safety Precautions

Follow all normal lab safety rules. In addition, take the following safety precautions:

1. Students with an allergy to yeast should notify the teacher immediately.

2. Put on sanitized indirectly vented chemical-splash goggles and laboratory apron and gloves before starting the lab activity.

3. Handle all glassware with care to avoid breakage. Sharp glass edges can cut skin!

4. Review the important information on chemicals on the safety data sheet, and use caution when handling chemicals.

5. Follow all safety rules that apply when working with electrical equipment, and use only GFCI-protected electrical receptacles.

6. Wash hands with soap and water after completing the lab activity.

Topics for the Explicit and Reflective Discussion

Concepts That Can Be Used to Justify the Evidence

To provide an adequate justification of their evidence, students must explain why they included the evidence in their arguments and make the assumptions underlying their analysis and interpretation of the data explicit. In this investigation, students can use the following concepts to help justify their evidence:

- The characteristics of a population that influence population growth
- How abiotic and biotic factors contribute to population growth
- How habitats provide the resources organisms need to survive
- Carrying capacity of an ecosystem for a species

We recommend that you review these concepts during the explicit and reflective discussion to help students make this connection.

How to Design Better Investigations

It is important for students to reflect on the strengths and weaknesses of the investigation they designed during the explicit and reflective discussion. Students should therefore be encouraged to discuss ways to eliminate potential flaws, measurement errors, or sources of bias in their investigations. To help students be more reflective about the design of their investigation, you can ask the following questions:

- What were some of the strengths of your investigation? What made it scientific?
- What were some of the weaknesses of your investigation? What made it less scientific?

- If you were to do this investigation again, what would you do to address the weaknesses in your investigation? What could you do to make it more scientific?

Crosscutting Concepts

This investigation is aligned with two crosscutting concepts found in *A Framework for K–12 Science Education,* and you should review these concepts during the explicit and reflective discussion.

- *Patterns:* Patterns are often used to guide the organization and classification of life on Earth in life science. In addition, a major objective in biology is to identify the underlying cause of observed patterns, such as the way in which many populations go through cycles of growth and decline over time.
- *Stability and change:* It is critical to understand what makes a system (such as an ecosystem) stable or unstable and what controls rates of change in a system.

The Nature of Science and the Nature of Scientific Inquiry

This investigation is aligned with two important concepts related to the *nature of science* (NOS) and the *nature of scientific inquiry* (NOSI), and you should review these concepts during the explicit and reflective discussion.

- *The difference between observations and inferences:* An observation is a descriptive statement about a natural phenomenon, whereas an inference is an interpretation of an observation. Students should also understand that current scientific knowledge and the perspectives of individual scientists guide both observations and inferences. Thus, different scientists can have different but equally valid interpretations of the same observations due to differences in their perspectives and background knowledge.
- *The nature and role of experiments:* Scientists use experiments to test the validity of a hypothesis (i.e., a tentative explanation) for an observed phenomenon. Experiments include a test and the formulation of predictions (expected results) if the test is conducted and the hypothesis is valid. The experiment is then carried out and the predictions are compared with the observed results of the experiment. If the predictions match the observed results, then the hypothesis is supported. If the predictions do not match the observed results, then the hypothesis is not supported. A signature feature of an experiment is the control of variables to help eliminate alternative explanations for observed results.

Hints for Implementing the Lab

- This investigation involves a lot of work with the microscope, so be sure to introduce students to the skills needed to use a microscope before starting this lab.

- If available, use a video microscope to demonstrate the counting procedure for the whole class.

- Have students select a group member to do the cell counts during spare time on the days where that is the only investigation activity required, but be sure that several group members get opportunities to collect cell count data.

- Emphasize the need for controlling variables when groups design their investigation; for example, they should control for environmental factors that are not being tested by that group.

Topic Connections

Table 9.2 provides an overview of the scientific practices, crosscutting concepts, disciplinary core ideas, and supporting ideas at the heart of this lab investigation. In addition, it lists NOS and NOSI concepts for the explicit and reflective discussion. Finally, it lists literacy and mathematics skills (*CCSS ELA* and *CCSS Mathematics*) that are addressed during the investigation.

TABLE 9.2

Lab 9 alignment with standards

Scientific practices	• Asking questions and defining problems • Developing and using models • Planning and carrying out investigations • Analyzing and interpreting data • Constructing explanations • Engaging in argument from evidence • Obtaining, evaluating, and communicating information
Crosscutting concepts	• Patterns • Stability and change
Core idea	• LS2: Ecosystems: Interactions, energy, and dynamics
Supporting ideas	• Populations • Population growth • Abiotic factors • Biotic factors • Carrying capacity
NOS and NOSI concepts	• Observations and inferences • Nature and role of experiments
Literacy connections (CCSS ELA)	• *Reading:* Key ideas and details, craft and structure, integration of knowledge and ideas • *Writing:* Text types and purposes, production and distribution of writing, research to build and present knowledge, range of writing • *Speaking and listening:* Comprehension and collaboration, presentation of knowledge and ideas
Mathematics connections (CCSS Mathematics)	• Make sense of problems and persevere in solving them • Reason abstractly and quantitatively • Construct viable arguments and critique the reasoning of others • Model with mathematics • Use appropriate tools strategically • Look for and express regularity in repeated reasoning

LAB 9

Lab 9. Population Growth: What Factors Limit the Size of a Population of Yeast?

Introduction

All populations of living things change in size over time. The human population is no different. A population is a group of individuals that belong to the same species and live in the same region at the same time. In 1800, the total number of people on Earth was about 1 billion. Now the world population is 7 billion. This observation has caused many scientists to ask questions such as "What caused this dramatic increase in the human population?" and "How long will the human population continue to grow?"

Human population growth has been a "hot topic" of discussion in the popular media and among ecologists (scientists who study how organisms interact with each other and the environment). To better understand human population growth on Earth, it is necessary to study population dynamics, or the changing size, density, and range of a population. Population dynamics is an area of life science that focuses on these changes. Scientists studying *population dynamics* investigate how different factors in living organisms and the environment shape those changes. However, humans are not the only living things studied through population dynamics. Factors related to all living organisms can include what they eat and drink, the amount of space they need, and how they interact with other organisms. Many of these factors are considered *biotic factors*, because they involve living things in an ecosystem. Factors related to the environment can include how much space is available and the weather and climate patterns in a particular area. Many of these factors are considered *abiotic factors*, because they involve nonliving pieces of an ecosystem.

By investigating these biotic and abiotic factors, scientists determine important relationships that help populations grow or reduce the number of organisms in an area. Scientists study the population dynamics of many different organisms to help us understand more about what affects human populations. Also, scientists study how human populations change other populations of organisms. Knowing about these relationships helps scientists and policy makers make decisions about how to manage natural resources and organisms in many different types of ecosystems.

Your Task

Design an investigation on how the size of a population of yeast changes over time in response to different factors such as *amount of food*, *amount of space*, and *the initial size of the population*. Yeasts are single-celled organisms in the Fungi kingdom. One species of yeast called *Saccharomyces cerevisiae* (Figure L9.1) has been used in baking and alcoholic beverages for thousands of years. Scientists have also used it to gather information about how cells function because it reproduces quickly. In fact, *S. cerevisiae* and many other species of yeast can produce a new generation every two hours. Therefore, a population of yeast could potentially increase in size very quickly if something did not prevent the size of the population from growing over time.

The guiding question of this investigation is, **What factors limit the size of a population of yeast?**

FIGURE L9.1

Microscopic image of yeast (*Saccharomyces cerevisiae*)

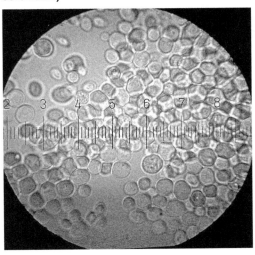

Materials

You may use any of the following materials during your investigation:

Consumables	Equipment
• Yeast culture	• Graduated pipette (1 ml)
• Sugar solution	• Test tubes
• Iodine solution	• Test tube rack
• Distilled water	• Simple compound light microscope
	• Microscope slides
	• Cover slips
	• Calculator
	• Sanitized indirectly vented chemical-splash goggles
	• Chemical-resistant apron
	• Gloves

Safety Precautions

Follow all normal lab safety rules. In addition, take the following safety precautions:

1. If you have any allergies to yeast, be sure to discuss this with your teacher immediately!

2. Put on sanitized indirectly vented chemical-splash goggles and laboratory apron and gloves before starting the lab activity.

3. Handle all glassware with care to avoid breakage. Sharp glass edges can cut skin!

LAB 9

4. Review the important information on chemicals on the safety data sheet, and use caution when handling chemicals.

5. Follow all safety rules that apply when working with electrical equipment, and use only GFCI-protected electrical receptacles.

6. Wash hands with soap and water after completing the lab activity.

Investigation Proposal Required?   ☐ Yes   ☐ No

Getting Started

Brainstorm with your group one possible factor that you think will limit the size of the yeast population. While designing your investigation, consider this variable while writing your hypotheses (e.g., population size is limited by the amount of food available, space available, and other factors). Be sure to include a control condition in your investigation, which will include environmental conditions that you will not change across different treatment conditions. A control condition usually represents the "normal" environment for the organism you are studying. You will need to track the size of a population of yeast for the next 72 hours.

Listed below are some important tips for working with yeast.

Setting up a population of yeast:

1. Use a graduated pipette to transfer 1 ml of the yeast from the class culture to a standard test tube. Measure carefully. In this case, more is not better.

2. Add two drops of iodine to yeast in the test tube. Be sure to drop the iodine into the culture, not on the side of the test tube. (The iodine will help you to see the cells under the microscope.)

3. Add water to the test tube (this can range from 1 ml to 5 ml depending on how much space you want to give the yeast population).

4. Add sugar solution to the test tube (this can range from 1 ml to 3 ml depending on how much food you want to give the yeast population).

Counting the number of yeast in your test tube:

1. Because yeast cells tend to settle out of solution, you will need to stir the yeast in your test tube so that the cells are evenly distributed. This must be done gently to avoid foaming the culture.

2. Use the 1 ml pipette to transfer 0.1 ml (a single drop) from the test tube to the graduated microscope slide.

3. Carefully lower a cover slip onto the drop to make a wet mount slide. Observe the slide under low power and identify the yeast cells.

4. Count the number of yeast cells in three different fields of view under high power. Select those fields of view from different areas of the slide.

5. Add the total number of cells you counted in all three squares and find the average number of cells per field of view.

To answer the guiding question, you must first determine what type of data you need to collect, how you will collect it, and how you will analyze it. To determine *what type of data you need to collect,* think about the following question:

- What type of measurements or observations will you need to record during your investigation?

To determine *how you will collect your data,* think about the following questions:

- What will serve as a control (or comparison) condition?
- What types of treatment conditions will you need to set up and how will you do it?
- During the experiment, when will you collect data and how often will you collect it?
- How will you make sure that your data are of high quality (i.e., how will you reduce error)?
- How will you keep track of the data you collect and how will you organize it?

To determine *how you will analyze your data,* think about the following questions:

- How will you determine if there is a difference between the treatment conditions and the control condition?
- How will you calculate change over time?
- What type of graph could you create to help make sense of your data?

Connections to Crosscutting Concepts, the Nature of Science, and the Nature of Scientific Inquiry

As you work through your investigation, be sure to think about

- how scientists try to identify patterns in nature to better understand it,
- how living things go through periods of stability followed by periods of change,
- the difference between observations and inferences in science, and
- the role of experiments in science

LAB 9

Initial Argument

Once your group has finished collecting and analyzing your data, you will need to develop an initial argument. Your argument must include a claim, evidence to support your claim, and a justification of the evidence. The claim is your group's answer to the guiding question. The evidence is an analysis and interpretation of your data. Finally, the justification of the evidence is why your group thinks the evidence matters. The justification of the evidence is important because scientists can use different kinds of evidence to support their claims. Your group will create your initial argument on a whiteboard. Your whiteboard should include all the information shown in Figure L9.2.

FIGURE L9.2

Argument presentation on a whiteboard

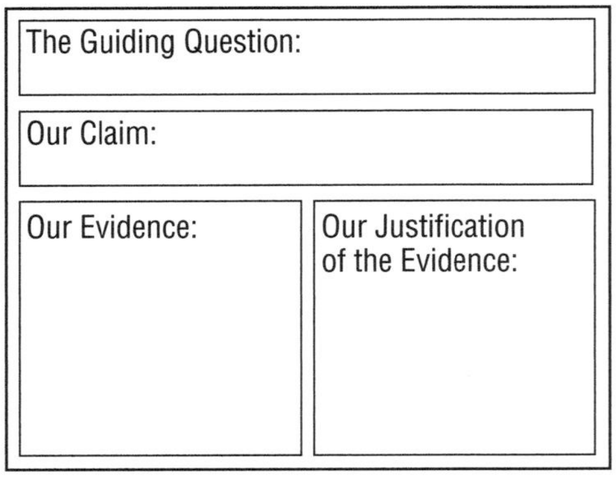

The Guiding Question:	
Our Claim:	
Our Evidence:	Our Justification of the Evidence:

Argumentation Session

The argumentation session allows all of the groups to share their arguments. One member of each group will stay at the lab station to share that group's argument, while the other members of the group go to the other lab stations one at a time to listen to and critique the arguments developed by their classmates. This is similar to how scientists present their arguments to other scientists at conferences. If you are responsible for critiquing your classmates' arguments, your goal is to look for mistakes so these mistakes can be fixed and they can make their argument better. The argumentation session is also a good time to think about ways you can make your initial argument better. Scientists must share and critique arguments like this to develop new ideas.

To critique an argument, you might need more information than what is included on the whiteboard. You will therefore need to ask the presenter lots of questions. Here are some good questions to ask:

- What did your group do to collect the data? Why do you think that way is the best way to do it?
- What did your group do to analyze the data? Why did your group decide to analyze it that way?
- What other ways of analyzing and interpreting the data did your group talk about?
- What did your group do to make sure that these calculations are correct?
- Why did your group decide to present your evidence in that way?
- What other claims did your group discuss before you decided on that one? Why did your group abandon those other ideas?
- How sure are you that your group's claim is accurate? What could you do to be more certain?

Once the argumentation session is complete, you will have a chance to meet with your group and revise your original argument. Your group might need to gather more data or design a way to test one or more alternative claims as part of this process. Remember, your goal at this stage of the investigation is to develop the most valid or acceptable answer to the research question!

Report

Once you have completed your research, you will need to prepare an investigation report that consists of three sections that provide answers to the following questions:

1. What question were you trying to answer and why?

2. What did you do during your investigation and why did you conduct your investigation in this way?

3. What is your argument?

Your report should answer these questions in two pages or less. The report must be typed, and any diagrams, figures, or tables should be embedded into the document. Be sure to write in a persuasive style; you are trying to convince others that your claim is acceptable or valid!

LAB 9

Lab 9. Population Growth: What Factors Limit the Size of a Population of Yeast?

1. Describe how the amount of food or amount of space available might influence the size of a population of organisms.

2. In a laboratory, yeast can be grown in a very controlled setting without many external influences on their population. In nature, however, conditions are constantly changing. Using what you know about population dynamics, describe what might happen to a population of organisms if the climate where they live changes.

3. *Observation* and *inference* are two words that mean the same thing.

    a. I agree with this statement.
    b. I disagree with this statement.

    Explain your answer, using an example from your investigation about population growth.

4. If scientists want to be certain about an idea, they must conduct an experiment to test it.

    a. I agree with this statement.

    b. I disagree with this statement.

    Explain your answer, using an example from your investigation about population growth.

5. Scientists often look for patterns when they are investigating the world. Explain why identifying patterns in the world is helpful, using an example from your investigation about population growth.

6. Understanding how living things go through periods of stability followed by periods of change is important for scientists. Explain why understanding the relationship between stability and change is important, using an example from your investigation about population growth.

LAB 10

Teacher Notes

Lab 10. Predator-Prey Relationships: How Is the Size of a Predator Population Related to the Size of a Prey Population?

Purpose

The purpose of this lab is to *introduce* students to the concept of a predator-prey population size relationship and some of the factors that affect population size and survival. This lab also gives students the opportunity to design and carry out an investigation using an online simulation. The simulation, called *Wolf Sheep Predation* (Wilensky 1997), was created using NetLogo, a multiagent programmable modeling environment developed at the Center for Connected Learning and Computer-Based Modeling at Northwestern University (Wilensky 1999). The simulation allows students to explore a simple model ecosystem that consists of wolves, sheep, and grass. In the simulation, wolves and sheep wander randomly around the landscape. When a wolf bumps into a sheep, the wolf eats the sheep and gains energy. If a wolf does not gain enough energy, it dies. The sheep must also eat grass to maintain their energy. If a sheep does not gain enough energy, it will also die.

This lab gives students an opportunity to explore the cause-and-effect relationships found in nature. Students will also learn about the role of models in science and how they can be used to study different systems of living things. Students will have the opportunity to reflect on how scientific knowledge can change over time and how scientists use different methods to investigate questions they have about the world.

The Content

Individuals are always part of a larger group of organisms from the same species, called a *population*. In some respects, populations act like individual organisms. They require space and nutrients. They have daily and seasonal cycles, including when they sleep and eat. They grow and die. The size of any population is in large part determined by a balance between several factors; some of these factors are obvious and some are not. Ultimately the biological success of a population is measured by its size over time. The factors that affect how well organisms grow and survive are the factors that determine population size.

Populations have unique properties, including an age structure, a death rate, and a birth rate. The *age structure* describes how organisms in a population are grouped in different age groups. A growing population will have a larger number of individuals in the younger groups than in the older groups; the reverse will be true of a dying population, with more individuals in the older groups than in the younger ones. The *birth rate* and *death rate* of a population tell us how many individual organisms are born into a population and die off in that population on a yearly basis. If a population has a higher birth rate than death rate, that population is growing. If a population's death rate is higher than its birth rate, that population is shrinking. These properties represent some of the ways scientists measure the health of a population. Knowing about these properties can help them study ways that will help bring species back from the brink of extinction or how other elements in the environment are affected by those organisms. For humans, these measures of population growth are used to make a variety of policy decisions, including projections of health care costs, the size of future labor forces and tax bases, and the value of certain public health initiatives.

A population of one kind of organism also interacts with other populations in the same area where it lives. Several populations interacting with each other form a *community*. Organisms from different populations but from the same community relate in several different ways. Some organisms from different populations help each other survive through providing resources for one another, which is known as *mutualism*. An example of this relationship would be bees and flowering plants. The bees find food when the flowers bloom on the plant, and that food helps the bee population survive. When bees visit multiple flowers, they also carry pollen from one flower to another. This movement of pollen helps the flowering plants to reproduce, so the flower population continues to survive.

Another kind of relationship found between populations is known as *predation*. Predation involves organisms from one population using organisms from another population as food. The organism that is used as food is called *prey*, and the organism that eats the other organism is called a *predator*. There are many examples of predator-prey relationships. When you pick vegetables from a garden, you are a predator and the plant is prey. Seagulls and bears are predators of several kinds of fish that are their prey. Predator-prey relationships are very common in different communities of organisms.

Some predators and prey are not closely related to each other, depending on the ecosystem. These "looser" relationships are typically due to multiple prey species being present in a specific community, so the predator does not rely on a sole food source. However, in a different ecosystem and community, the same predator-prey relationship could be "tighter" because there is less variety in the prey species available there. When predators and prey are more closely related like this, distinct trends in growth of the predator and prey populations typically result. As the prey relationship grows in size, there is more food available to the predator population. With more food, the predators are better able to survive and reproduce, so the predator population begins to grow. However, as the predator population grows, so does the demand for food. Eventually, there will be so many predators that the prey species cannot reproduce fast enough before being eaten. When this happens, the prey population will decrease. As that decrease continues, the predator population will lack the food it needs to survive. As the prey population shrinks, the predator population will also soon decrease. Eventually, the predator population decreases enough that the prey population can begin reproducing faster than it is eaten and will begin growing again. As the prey population begins growing, then so will the

LAB 10

FIGURE 10.1

Population size chart demonstrating boom-and-bust cycles for a predator-prey pair of populations

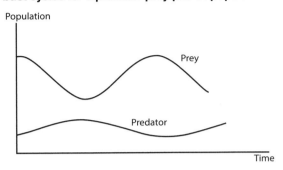

predator population. This growth and shrinking pattern in "tighter" predator-prey relationships is commonly called *boom-and-bust cycles* (Figure 10.1).

Timeline

The instructional time needed to implement this lab investigation is 130–200 minutes. Appendix 2 (p. 355) provides options for implementing this lab investigation over several class periods. Option C (200 minutes) should be used if students are unfamiliar with scientific writing, because this option provides extra instructional time for scaffolding the writing process. You can scaffold the writing process by modeling, providing examples, and providing hints as students write each section of the report. Option D (130 minutes) should be used if students are familiar with scientific writing and have the skills needed to write an investigation report on their own. In option D, students complete stage 6 (writing the investigation report) and stage 8 (revising the investigation report) as homework.

Materials and Preparation

The materials needed to implement this investigation are listed in Table 10.1. Wilensky's *Wolf Sheep Predation* simulation, available at *http://ccl.northwestern.edu/netlogo/models/WolfSheepPredation*, is free to use and can be downloaded as part of an application. You should access the website and learn how the simulation works before beginning the lab investigation. In addition, it is important to check if students can access and use the simulation from a school computer, because some schools have set up firewalls and other restrictions on web browsing. Be sure to work with your instructional technology staff to download the NetLogo application.

TABLE 10.1

Materials list

Item	Quantity
Computer with NetLogo application	1 per group
Investigation Proposal A (optional)	1 per group
Whiteboard, 2' × 3'*	1 per group
Lab Handout	1 per student
Peer-review guide	1 per student
Checkout Questions	1 per student

* As an alternative, students can use computer and presentation software such as Microsoft PowerPoint or Apple Keynote to create their arguments.

Safety Precautions

Follow all normal lab safety rules.

Topics for the Explicit and Reflective Discussion

Concepts That Can Be Used to Justify the Evidence

To provide an adequate justification of their evidence, students must explain why they included the evidence in their arguments and make the assumptions underlying their analysis and interpretation of the data explicit. In this investigation, students can use the following concepts to help justify their evidence:

- How populations interact with each other in different ways
- How predators rely on other organisms for food, which are called prey
- The characteristics of a population that influence population growth, including the number of prey available in an area for a predator to eat
- Boom-and-bust patterns, which occur in predator-prey relationships that have less variation in available resources

We recommend that you review these concepts during the explicit and reflective discussion to help students make this connection.

How to Design Better Investigations

It is important for students to reflect on the strengths and weaknesses of the investigation they designed during the explicit and reflective discussion. Students should therefore be encouraged to discuss ways to eliminate potential flaws, measurement errors, or sources of bias in their investigations. To help students be more reflective about the design of their investigation, you can ask the following questions:

- What were some of the strengths of your investigation? What made it scientific?
- What were some of the weaknesses of your investigation? What made it less scientific?
- If you were to do this investigation again, what would you do to address the weaknesses in your investigation? What could you do to make it more scientific?

Crosscutting Concepts

This investigation is aligned with two crosscutting concepts found in *A Framework for K–12 Science Education*, and you should review these concepts during the explicit and reflective discussion.

LAB 10

- *Cause and effect: Mechanism and explanation:* Natural phenomena have causes, and uncovering causal relationships (e.g., how changes in environmental conditions affect other elements in the environment) is a major activity of science.

- *Systems and system models:* It is critical for scientists to be able to define the system under study (e.g., the components of a habitat) and then make a model of it to understand it. Models can be physical, conceptual, or mathematical.

The Nature of Science and the Nature of Scientific Inquiry

This investigation is aligned with two important concepts related to the *nature of science* (NOS) and the *nature of scientific inquiry* (NOSI), and you should review these concepts during the explicit and reflective discussion.

- *Changes in scientific knowledge over time:* A person can have confidence in the validity of scientific knowledge but must also accept that scientific knowledge may be abandoned or modified in light of new evidence or because existing evidence has been reconceptualized by scientists. There are many examples in the history of science of both evolutionary changes (i.e., the slow or gradual refinement of ideas) and revolutionary changes (i.e., the rapid abandonment of a well-established idea) in scientific knowledge.

- *Methods used in scientific investigations:* Examples of methods include experiments, systematic observations of a phenomenon, literature reviews, and analysis of existing data sets; the choice of method depends on the objectives of the research. There is no universal step-by step scientific method that all scientists follow; rather, different scientific disciplines (e.g., chemistry vs. biology) and fields within a discipline (e.g., ecology vs. molecular biology) use different types of methods, use different core theories, and rely on different standards to develop scientific knowledge.

Hints for Implementing the Lab

- Allow the students to play with the simulation as part of the tool talk before they begin to design their investigation or fill out an investigation proposal. This gives students a chance to see what they can and cannot do with the simulation.

- To help focus students, establish fixed values for some variables (i.e., grass present so rabbits can eat; show quantitative values), but be sure to discuss the reasoning for setting those values a certain way.

- There are several factors that can be adjusted in this simulation. For simplicity and focus on predator-prey relationships, you will probably not want to adjust too many of these factors. A whole-class discussion orienting students to these features will be helpful and provide an opportunity to determine fixed states for some of the factors.

- The simulation will kill off one of the populations eventually, and this characteristic can set the stage to discuss the limitations of simulations and models.

Topic Connections

Table 10.2 provides an overview of the scientific practices, crosscutting concepts, disciplinary core ideas, and supporting ideas at the heart of this lab investigation. In addition, it lists NOS and NOSI concepts for the explicit and reflective discussion. Finally, it lists literacy and mathematics skills (*CCSS ELA* and *CCSS Mathematics*) that are addressed during the investigation.

TABLE 10.2

Lab 10 alignment with standards

Scientific practices	• Asking questions and defining problems • Developing and using models • Planning and carrying out investigations • Analyzing and interpreting data • Using mathematics and computational thinking • Constructing explanations • Engaging in argument from evidence • Obtaining, evaluating, and communicating information
Crosscutting concepts	• Cause and effect: Mechanism and explanation • Systems and system models
Core idea	• LS2: Ecosystems: Interactions, energy, and dynamics
Supporting ideas	• Populations • Population growth • Community • Predator • Prey • Boom-and-bust cycles
NOS and NOSI concepts	• Changes in scientific knowledge over time • Methods used in scientific investigations
Literacy connections (CCSS ELA)	• *Reading:* Key ideas and details, craft and structure, integration of knowledge and ideas • *Writing:* Text types and purposes, production and distribution of writing, research to build and present knowledge, range of writing • *Speaking and listening:* Comprehension and collaboration, presentation of knowledge and ideas
Mathematics connections (CCSS Mathematics)	• Make sense of problems and persevere in solving them • Reason abstractly and quantitatively • Construct viable arguments and critique the reasoning of others • Model with mathematics • Use appropriate tools strategically

LAB 10

References

Wilensky, U. 1997. NetLogo Wolf Sheep Predation model. Evanston, IL: Center for Connected Learning and Computer-Based Modeling, Northwestern Institute on Complex Systems, Northwestern University. *http://ccl.northwestern.edu/netlogo/models/WolfSheepPredation.*

Wilensky, U. 1999. NetLogo. Evanston, IL: Center for Connected Learning and Computer-Based Modeling, Northwestern Institute on Complex Systems, Northwestern University. *http://ccl.northwestern.edu/netlogo.*

Lab Handout

Lab 10. Predator-Prey Relationships: How Is the Size of a Predator Population Related to the Size of a Prey Population?

Introduction

John Donne wrote, "No man is an island." The same is true for any individual plant or animal. Individuals are always part of a larger group of organisms from the same species, called a *population*. In some respects, populations act like individual organisms. They require space and nutrients. They have daily and seasonal cycles, including when they sleep and eat. They grow and die. The size of any population is in large part determined by a balance between several factors; some of these factors are obvious and some are not. Ultimately the biological success of a population is measured by its size over time. The factors that affect how well organisms grow and survive are the factors that determine population size.

A population of one kind of organism also interacts with other populations in the same area where it lives. Several populations interacting with each other form a *community*. Organisms from different populations but from the same community relate in several different ways. Some organisms from different populations help each other survive through providing resources for one another, which is known as *mutualism*. An example of this relationship would be bees and flowering plants. The bees find food when the flowers bloom on the plant, and that food helps the bee population survive. When bees visit multiple flowers, they also carry pollen from one flower to another. This movement of pollen helps the flowering plants to reproduce, so the flower population continues to survive.

Another kind of relationship found between populations is known as *predation*. Predation involves organisms from one population using organisms from another population as food. The organism that is used as food is called *prey*, and the organism that eats the other organism is called a *predator*. There are many examples of predator-prey relationships. When you pick vegetables from a garden, you are a predator and the plant is prey. Seagulls and bears are predators of several kinds of fish that are their prey. Predator-prey relationships are very common in different communities of organisms.

Your Task

Using a computer simulation, investigate how a population of a predator (wolves) and a population of its prey (sheep) interact with each other and the local environment over time.

The guiding question of this investigation is, **How is the size of a predator population related to the size of a prey population?**

LAB 10

Materials

You will use an online simulation called *Wolf Sheep Predation* to conduct your investigation. You can find the simulation in the module library of the NetLogo program on your lab computer: *http://ccl.northwestern. edu/netlogo/models/WolfSheepPredation*.

Safety Precautions

Follow all normal lab safety rules.

Investigation Proposal Required?   ☐ Yes   ☐ No

Getting Started

The *Wolf Sheep Predation* simulation allows you to explore how populations of predators and prey interact with each other over time. This simulation is designed to follow the rules of nature so you can use it to see what happens to the size of a population when you change different environmental factors, such as size of initial population and availability of food for the prey species.

In the simulation, wolves and sheep wander randomly around the landscape. When a wolf bumps into a sheep, the wolf eats the sheep. Each step costs the wolves energy, and they must eat sheep to replenish their energy; when they run out of energy, they die. Each wolf or sheep reproduces at a constant rate. You can also choose to include grass in addition to wolves and sheep in the simulation. If you add the grass, the sheep must eat grass to maintain their energy; when they run out of energy, they die. Once grass is eaten, it will only regrow after a fixed amount of time.

You can change a wide range of factors in the simulation:

- INITIAL-NUMBER-SHEEP: the initial size of the sheep population
- INITIAL-NUMBER-WOLVES: the initial size of the wolf population
- WOLF-GAIN-FROM-FOOD: the amount of energy a wolf gets for every sheep eaten
- SHEEP-REPRODUCE: how often sheep reproduce
- WOLF-REPRODUCE: how often wolves reproduce

To answer the guiding question, you must determine what type of data you need to collect, how you will collect it, and how you will analyze it. To determine *what type of data you need to collect*, think about the following questions:

- How will you determine if the composition of the sheep and wolf populations changes over time?

- What will serve as your dependent variable?
- What type of measurements or observations will you need to record during your investigation?

To determine *how you will collect your data*, think about the following questions:

- What will serve as a control condition?
- What types of treatment conditions will you need to set up and how will you do it?
- How many trials will you need to conduct?
- How long will you need to run the simulation during each trial?
- How often will you collect data and when will you do it?
- How will you keep track of the data you collect and how will you organize it?

To determine *how you will analyze your data*, think about the following questions:

- How will you determine if there is a difference between the different treatment conditions and the control condition?
- What type of calculations will you need to make?
- What type of graph could you create to help make sense of your data?

Connections to Crosscutting Concepts, the Nature of Science, and the Nature of Scientific Inquiry

As you work through your investigation, be sure to think about

- the importance of understanding cause-and-effect relationships for natural phenomena,
- the use of models to study systems,
- the way scientific knowledge can change over time, and
- the different types of methods that scientists use to answer questions.

Initial Argument

Once your group has finished collecting and analyzing your data, you will need to develop an initial argument. Your argument must include a claim, evidence to support your claim, and a justification of the evidence. The claim is your group's answer to the guiding question. The evidence is an analysis and interpretation of your data. Finally, the justification of the evidence is why your group thinks the evidence matters. The justification of the evidence is important because scientists can use different kinds of evidence to support their claims. Your group will create your initial argument on a whiteboard. Your whiteboard should include all the information shown in Figure L10.1 (p. 176).

LAB 10

Argument presentation on a whiteboard

The Guiding Question:	
Our Claim:	
Our Evidence:	Our Justification of the Evidence:

Argumentation Session

The argumentation session allows all of the groups to share their arguments. One member of each group will stay at the lab station to share that group's argument, while the other members of the group go to the other lab stations one at a time to listen to and critique the arguments developed by their classmates. This is similar to how scientists present their arguments to other scientists at conferences. If you are responsible for critiquing your classmates' arguments, your goal is to look for mistakes so these mistakes can be fixed and they can make their argument better. The argumentation session is also a good time to think about ways you can make your initial argument better. Scientists must share and critique arguments like this to develop new ideas.

To critique an argument, you might need more information than what is included on the whiteboard. You will therefore need to ask the presenter lots of questions. Here are some good questions to ask:

- What did your group do to collect the data? Why do you think that way is the best way to do it?
- What did your group do to analyze the data? Why did your group decide to analyze it that way?
- What other ways of analyzing and interpreting the data did your group talk about?
- Why did your group decide to present your evidence in that way?
- What other claims did your group discuss before you decided on that one? Why did your group abandon those other ideas?
- How sure are you that your group's claim is accurate? What could you do to be more certain?

Once the argumentation session is complete, you will have a chance to meet with your group and revise your original argument. Your group might need to gather more data or design a way to test one or more alternative claims as part of this process. Remember, your goal at this stage of the investigation is to develop the most valid or acceptable answer to the research question!

Report

Once you have completed your research, you will need to prepare an investigation report that consists of three sections that provide answers to the following questions:

1. What question were you trying to answer and why?

2. What did you do during your investigation and why did you conduct your investigation in this way?

3. What is your argument?

Your report should answer these questions in two pages or less. The report must be typed, and any diagrams, figures, or tables should be embedded into the document. Be sure to write in a persuasive style; you are trying to convince others that your claim is acceptable or valid!

LAB 10

Lab 10. Predator-Prey Relationships: How Is the Size of a Predator Population Related to the Size of a Prey Population?

1. In some ecosystems there may be multiple predators. Using what you know about predator-prey relationships, describe how it is possible for multiple predators to exist in the same habitat.

2. In an ecosystem, the size of the predator population is related to the size of the prey population. Describe what would happen if the predator population reproduced faster than the prey population.

3. Scientific knowledge changes so quickly that it should be considered unstable.

    a. I agree with this statement.

    b. I disagree with this statement.

    Explain your answer, using an example from your investigation about predator-prey relationships.

4. It is important for scientists to use a variety of methods to learn about the natural world.

   a. I agree with this statement.
   b. I disagree with this statement.

   Explain your answer, using an example from your investigation about predator-prey relationships.

5. Scientists often study potential cause-and-effect relationships when they investigate the natural world. Explain why it is important to understand causes and effects, using an example from your investigation about predator-prey relationships.

6. Scientists develop models to help them understand the natural world. Sometimes the models scientists develop are similar to the computer model that you used in the predator-prey investigation. Explain why models are helpful, using an example from your investigation about predator-prey relationships.

Application Labs

LAB 11

Lab 11. Food Webs and Ecosystems: Which Member of an Ecosystem Would Affect the Food Web the Most If Removed?

Purpose

The purpose of this lab is for students to *apply* their understanding of ecosystems and the eating relationships that occur within them. Specifically, students must consider the removal of different species from an ecosystem by humans and which removals would cause the most damage to the overall food web. This lab gives students an opportunity to learn about the role of models in science and how they can be used to study different systems of living things. Students will also explore the flow of energy and matter through systems. Students will have the opportunity to reflect on how science is connected to the society and culture in which it is practiced and how imagination and creativity are important to science.

The Content

An *ecosystem* includes all the living and nonliving pieces of a particular area of the planet. The living parts of an ecosystem include the plants, insects, bacteria, mammals, birds, and other living organisms present in an area. The nonliving pieces of an ecosystem include the rocks, water sources, gases in the air, and constructed structures like buildings and roads. All of the pieces of a specific ecosystem will connect and work together in different ways so the living members can try to survive in that ecosystem. An important set of relationships in an ecosystem involves eating to survive. How living things eat in an ecosystem is important to understand because eating involves the transfer of energy.

Living things in an ecosystem must eat other living things in the ecosystem to get the energy they need to survive. The only organisms that do not have to eat other organisms for their energy are called *producers*. Producers are organisms that create their own food by harvesting energy from other sources, such as the Sun. Plants are the most common type of producers found in an ecosystem. Bacteria can also be considered producers, especially if they use chlorophyll to harvest solar energy in a manner similar to plants. Another term used to describe producers is *autotrophs*, which are organisms that produce the organic molecules they need to survive using light or chemical energy. Producers in some ecosystems, such as deep-sea hydrothermal vent communities, use chemicals instead of light to produce the energy and matter they need to survive.

If an organism is not a producer in an ecosystem, then it is considered a *consumer*. Consumers are organisms that have to eat other living things to get the energy they need to survive. Some consumers will eat only the plants in an ecosystem, some consumers will eat only other consumers, and still other consumers will eat both the plants and other consumers. All consumers can also be referred to as *heterotrophs*, because they have to meet their energy and matter requirements for living through sources other than themselves.

Different organisms have different energy needs, which will influence what food they eat. In any ecosystem, there can be multiple producers and types of consumers. One way that scientists try to understand these relationships in an ecosystem is through designing *food webs*. A food web is a diagram that models the feeding relationships in an ecosystem. It can also be considered the combination of all the unique *food chains* present in an ecosystem. Food chains are models that represent the sequential eating relationship among a group of organisms present in an ecosystem. There can be many food chains present in a single ecosystem. One species of organism can be involved in multiple food chains. Food webs help show all the individual food chains operating in an ecosystem and how they overlap.

FIGURE 11.1 _____

Example of a food web diagram, showing the eating relationships in an ecosystem

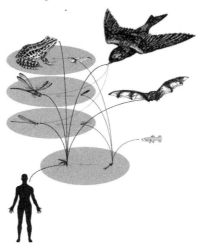

Figure 11.1 provides an example of a food web. Notice how each organism has some lines with arrows pointed into it and other lines with arrows coming out of it. A line with an arrow coming out of an organism indicates what that organism eats; in contrast, a line with an arrow pointing into an organism indicates that the organism is eaten by the organism at the other end of the line. Each oval represents the young and adult forms of an organism. Some animals will only eat the young version (e.g., eggs or larvae) of another organism. Also notice how one type of organism in the food web can be a food source for several other organisms in the same ecosystem.

In a food web, consumers can be ordered based upon their relative position in their respective food chains and the overall food web. Consumers that feed only on the producers in the ecosystem are called *primary consumers*. Consumers that typically feed on primary consumers are considered *secondary consumers*. Based on the complexity of the food web (i.e., the number of different species in a specific ecosystem and their relevant food chains), consumers can also be classified as tertiary or quaternary. At the "top" of most food webs, there is a species that is considered the *keystone predator*. The keystone predator is an organism that does not have any competition for prey and regulates the population levels of the other larger predator species present in an ecosystem. Often when a keystone predator is removed from an ecosystem, the balance is upset and competition between top-level consumers can cause the ecosystem to fail.

The level of consumer that an organism is also influences how much energy it can access in an ecosystem. The amount of energy available to an ecosystem is determined by the number of producer species present. As energy transfers from one level of the food web to another (or up one "link" in a specific food chain), only 10% of the total energy available from the lower level becomes usable energy at the next higher level. This 10% energy principle for ecosystems helps explain why some species are more prevalent in certain ecosystems than others. Primary consumers have access to much more energy because they eat the producers directly and, thus, can support higher population levels in that ecosystem. On the other hand, high-level consumers, such as keystone predators, have to eat a lot more to meet their energy needs because their main prey also have to eat a lot more to meet their energy needs. The ability of a species to have several sources of food for energy also influences how large a population of that species an ecosystem can support. The ability of a species to use several food sources (and be involved in multiple food chains) is an important concept that this investigation will help students explore.

By understanding the food web of a certain ecosystem, scientists can also understand the impact human activity can have on that ecosystem. There are many situations in which humans try to remove a certain type of organism from an ecosystem, often for reasons involving public health or managing resources. Humans can add chemicals to an ecosystem that can get rid of certain plants or insects from an ecosystem. They can also hunt larger organisms that may be a higher-level organism in an ecosystem's food web. However, eliminating one type of organism from an ecosystem will have an impact on other organisms in that system. The scenario used to frame this investigation in the "Your Task" section of the Lab Handout reflects this tension and provides students with a real-world connection to the guiding question.

Timeline

The instructional time needed to implement this lab investigation is 130–200 minutes. Appendix 2 (p. 355) provides options for implementing this lab investigation over several class periods. Option C (200 minutes) should be used if students are unfamiliar with scientific writing, because this option provides extra instructional time for scaffolding the writing process. You can scaffold the writing process by modeling, providing examples, and providing hints as students write each section of the report. Option D (130 minutes) should be used if students are familiar with scientific writing and have the skills needed to write an investigation report on their own. In option D, students complete stage 6 (writing the investigation report) and stage 8 (revising the investigation report) as homework.

Materials and Preparation

The materials needed to implement this investigation are listed in Table 11.1. The slides can be found at *www.nsta.org/publications/press/extras/adi-lifescience.aspx*.

TABLE 11.1
Materials list

Item	Quantity
Slides of marsh ecosystem organisms	1 per group
Investigation Proposal A (optional)	1 per group
Whiteboard, 2' × 3'*	1 per group
Lab Handout	1 per student
Peer-review guide	1 per student
Checkout Questions	1 per student

* As an alternative, students can use computer and presentation software such as Microsoft PowerPoint or Apple Keynote to create their arguments.

Safety Precautions

Follow all normal lab safety rules.

Topics for the Explicit and Reflective Discussion

Concepts That Can Be Used to Justify the Evidence

To provide an adequate justification of their evidence, students must explain why they included the evidence in their arguments and make the assumptions underlying their analysis and interpretation of the data explicit. In this investigation, students can use the following concepts to help justify their evidence:

- Food webs model the many feeding interactions that occur in an ecosystem.
- Producers and consumers have different roles in an ecosystem food web, based on their relationships to other organisms.
- Although one food source may be removed from an ecosystem, many species have multiple food sources, which can limit the change to the overall structure of the food web.

We recommend that you review these concepts during the explicit and reflective discussion to help students make this connection.

How to Design Better Investigations

It is important for students to reflect on the strengths and weaknesses of the investigation they designed during the explicit and reflective discussion. Students should therefore be encouraged to discuss ways to eliminate potential flaws, measurement errors, or sources of bias in their investigations. To help students be more reflective about the design of their investigation, you can ask the following questions:

- What were some of the strengths of your investigation? What made it scientific?

- What were some of the weaknesses of your investigation? What made it less scientific?

- If you were to do this investigation again, what would you do to address the weaknesses in your investigation? What could you do to make it more scientific?

Crosscutting Concepts

This investigation is aligned with two crosscutting concepts found in *A Framework for K–12 Science Education*, and you should review these concepts during the explicit and reflective discussion.

- *Systems and system models:* It is critical for scientists to be able to define the system under study (e.g., the components of an ecosystem) and then make a model of it to understand it. Models can be physical, conceptual, or mathematical.

- *Energy and matter: Flows, cycles, and conservation:* In science it is important to track how energy and matter move into, out of, and within systems.

The Nature of Science and the Nature of Scientific Inquiry

This investigation is aligned with two important concepts related to the *nature of science* (NOS) and the *nature of scientific inquiry* (NOSI), and you should review these concepts during the explicit and reflective discussion.

- *The influence of society and culture on science:* Science is influenced by the society and culture in which it is practiced because science is a human endeavor. Cultural values and expectations determine what scientists choose to investigate, how investigations are conducted, how research findings are interpreted, and what people see as implications. People also view some research as being more important than others because of cultural values and current events.

- *The importance of imagination and creativity in science:* Students should learn that developing explanations for or models of natural phenomena and then figuring out how they can be put to the test of reality is as creative as writing poetry, composing music, or designing skyscrapers. Scientists must also use their imagination and creativity to figure out new ways to test ideas and collect or analyze data.

Hints for Implementing the Lab

- Having paper copies of the organism slides can help students physically lay out their food webs.

- Laptop computers can be used for each group to view the slides if paper copies are not available.

- Help students focus on ways of quantifying the changes that occur in their different food web scenarios, which can help give them another piece of evidence.

Topic Connections

Table 11.2 provides an overview of the scientific practices, crosscutting concepts, disciplinary core ideas, and supporting ideas at the heart of this lab investigation. In addition, it lists NOS and NOSI concepts for the explicit and reflective discussion. Finally, it lists literacy and mathematics skills (*CCSS ELA* and *CCSS Mathematics*) that are addressed during the investigation.

TABLE 11.2

Lab 11 alignment with standards

Scientific practices	• Asking questions and defining problems • Developing and using models • Planning and carrying out investigations • Analyzing and interpreting data • Constructing explanations • Engaging in argument from evidence • Obtaining, evaluating, and communicating information
Crosscutting concepts	• Systems and system models • Energy and matter: Flows, cycles, and conservation
Core idea	• LS2: Ecosystems: Interactions, energy, and dynamics
Supporting ideas	• Food web • Producer • Consumer • Multiple sources of food
NOS and NOSI concepts	• Social and cultural influences • Imagination and creativity in science
Literacy connections (*CCSS ELA*)	• *Reading:* Key ideas and details, craft and structure, integration of knowledge and ideas • *Writing:* Text types and purposes, production and distribution of writing, research to build and present knowledge, range of writing • *Speaking and listening:* Comprehension and collaboration, presentation of knowledge and ideas
Mathematics connections (*CCSS Mathematics*)	• Make sense of problems and persevere in solving them • Construct viable arguments and critique the reasoning of others • Use appropriate tools strategically

LAB 11

Lab Handout

Lab 11. Food Webs and Ecosystems: Which Member of an Ecosystem Would Affect the Food Web the Most If Removed?

Introduction

An *ecosystem* includes all the living and nonliving pieces of a particular area of the planet. Living things in an ecosystem must eat other living things in the ecosystem to get the energy they need to survive. The only organisms that do not have to eat other organisms for their energy are called *producers*. Producers are organisms that create their own food by harvesting energy from other sources, such as the Sun. Plants are the most common type of producers found in an ecosystem. If an organism is not a producer in an ecosystem, then it is considered a *consumer*. Consumers are organisms that have to eat other living things to get the energy they need to survive. Some consumers will eat only the plants in an ecosystem, some consumers will eat only other consumers, and still other consumers will eat both the plants and other consumers.

Different organisms have different energy needs, which will influence what food they eat. In any ecosystem, there can be multiple producers and types of consumers. One way that scientists try to understand these relationships in an ecosystem is through designing *food webs*. A food web is a diagram that models the feeding relationships in an ecosystem. It can also be considered the combination of all the unique *food chains* present in an ecosystem. Food chains are models that represent the eating relationship among a group of organisms present in an ecosystem. There can be many food chains present in a single ecosystem. One species of organism can be involved in multiple food chains. Food webs help show all the individual food chains operating in an ecosystem and how they overlap.

Figure L11.1 provides an example of a food web. Notice how each organism has line arrows pointed into them and other line arrows coming out from them. A line with an arrow coming out of an organism indicates what that organism eats; in contrast, a line with an arrow pointing into an organism indicates that the organism is eaten by the organism at the other end of the line. Also notice how one type of organism in the food web can be a food source for several other organisms in the same ecosystem.

FIGURE L11.1

Example of a food web diagram, showing the eating relationships in an ecosystem

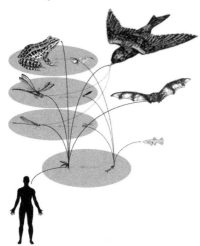

By understanding the food web of a certain ecosystem, scientists can also understand the impact human activity can have on that ecosystem. There are many situations in which humans try to remove a certain type of organism from an ecosystem, often for reasons involving public health or managing resources. Humans can add chemicals to an ecosystem that can get rid of certain plants or insects from an ecosystem. They can also hunt larger organisms that may be a higher-level organism in an ecosystem's food web. However, eliminating one type of organism from an ecosystem will have an impact on other organisms in that system.

Your Task

Explore the different roles of organisms in a specific ecosystem. A town has to decide which organism it should remove from its local ecosystem, which includes a swampy marsh. Many residents are worried about the mosquitoes that heavily populate the marsh. Others are concerned with the growth in algae and other weedlike plants in the marsh. Still other residents believe that the ducks in the marsh are a problem and should be hunted. Removing any one of these organisms, or others present in the marsh, will change the food web of the ecosystem. Your investigation should determine which organisms the town should remove to limit the amount of change to the existing food web.

The guiding question of this investigation is, **Which member of an ecosystem would affect the food web the most if removed?**

Materials

You will use slides of marsh ecosystem organisms during your investigation.

Safety Precautions

Follow all normal lab safety rules.

Investigation Proposal Required?   ☐ Yes   ☐ No

Getting Started

Your teacher can provide you with a copy of slides that have information about the different organisms in the marsh ecosystem. Use these slides to analyze what changes might occur to the original food web for the marsh when any one of the organisms is removed from it.

To answer the guiding question, you must determine what type of data you need to collect, how you will collect it, and how you will analyze it. To determine *what type of data you need to collect*, think about the following questions:

- What information on the slides relates most to the food web of the marsh?
- How will you represent the data you use in different ways?

LAB 11

- What type of measurements or observations will you need to record during your investigation?

To determine *how you will analyze your data*, think about the following questions:

- How will you understand what the current food web looks like?
- Do you need to analyze all the different organisms, or should you focus on types of organisms?
- What type of graph could you create to help make sense of your data?

Connections to Crosscutting Concepts, the Nature of Science, and the Nature of Scientific Inquiry

As you work through your investigation, be sure to think about

- the use of models to study systems,
- how tracking the flow of energy and matter through systems allows scientists to understand these systems,
- how science is influenced by society, and
- the role of imagination and creativity when solving problems in science.

Initial Argument

Once your group has finished collecting and analyzing your data, you will need to develop an initial argument. Your argument must include a claim, evidence to support your claim, and a justification of the evidence. The claim is your group's answer to the guiding question. The evidence is an analysis and interpretation of your data. Finally, the justification of the evidence is why your group thinks the evidence matters. The justification of the evidence is important because scientists can use different kinds of evidence to support their claims. Your group will create your initial argument on a whiteboard. Your whiteboard should include all the information shown in Figure L11.2.

FIGURE L11.2

Argument presentation on a whiteboard

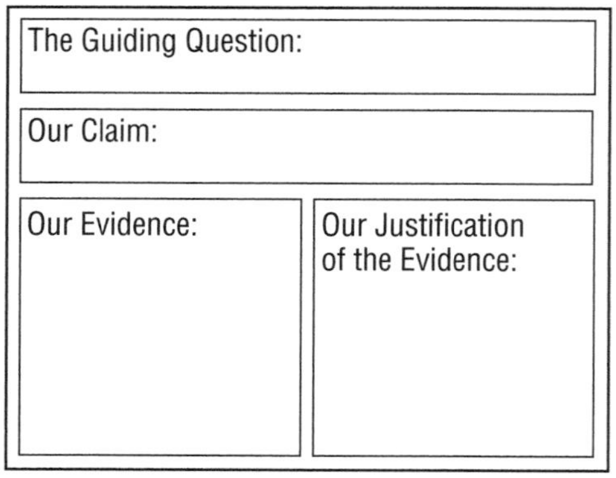

The Guiding Question:	
Our Claim:	
Our Evidence:	Our Justification of the Evidence:

Argumentation Session

The argumentation session allows all of the groups to share their arguments. One member of each group will stay at the lab station to share that group's argument, while the other members of the group go to the other lab stations one at a time to listen to and critique the arguments developed by their classmates. This is similar to how scientists present their arguments to other scientists at conferences. If you are responsible for critiquing your classmates' arguments, your goal is to look for mistakes so these mistakes can be fixed and they can make their argument better. The argumentation session is also a good time to think about ways you can make your initial argument better. Scientists must share and critique arguments like this to develop new ideas.

To critique an argument, you might need more information than what is included on the whiteboard. You will therefore need to ask the presenter lots of questions. Here are some good questions to ask:

- What did your group do to collect the data? Why do you think that way is the best way to do it?
- What did your group do to analyze the data? Why did your group decide to analyze it that way?
- What other ways of analyzing and interpreting the data did your group talk about?
- Why did your group decide to present your evidence in that way?
- What other claims did your group discuss before you decided on that one? Why did your group abandon those other ideas?
- How sure are you that your group's claim is accurate? What could you do to be more certain?

Once the argumentation session is complete, you will have a chance to meet with your group and revise your original argument. Your group might need to gather more data or design a way to test one or more alternative claims as part of this process. Remember, your goal at this stage of the investigation is to develop the most valid or acceptable answer to the research question!

Report

Once you have completed your research, you will need to prepare an investigation report that consists of three sections that provide answers to the following questions:

1. What question were you trying to answer and why?
2. What did you do during your investigation and why did you conduct your investigation in this way?
3. What is your argument?

Your report should answer these questions in two pages or less. The report must be typed, and any diagrams, figures, or tables should be embedded into the document. Be sure to write in a persuasive style; you are trying to convince others that your claim is acceptable or valid!

LAB 11

Lab 11. Food Webs and Ecosystems: Which Member of an Ecosystem Would Affect the Food Web the Most If Removed?

1. Imagine an ecosystem where mice eat the grass and foxes eat the mice. Explain what will happen to the population of foxes if there is a severe drought and all the grass dies.

2. The images below represent two food webs from different ecosystems that have similar animals. Which population of foxes would be least impacted by a drought that caused the grass to die in their ecosystem? Explain your reasoning.

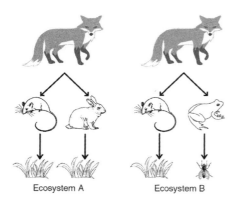

Ecosystem A   Ecosystem B

3. Society and culture often influence how scientists go about their work.

   a. I agree with this statement.

   b. I disagree with this statement.

Explain your answer, using an example from your investigation about food webs and ecosystems.

4. Scientists are very creative when they investigate the natural world.

   a. I agree with this statement.

   b. I disagree with this statement.

   Explain your answer, using an example from your investigation about food webs and ecosystems.

5. Scientists develop models to help them understand the natural world. Explain how a food web acts like a model and explain why such a model would be useful to scientists.

6. When scientists study the natural world, they often need to keep track of how matter and energy move through a system. Explain why understanding the flow of matter and energy is important, using an example from your investigation about food webs and ecosystems.

LAB 12

Teacher Notes

Lab 12. Matter in Ecosystems: How Healthy Are Your Local Ecosystems?

Purpose

The purpose of this lab is to have students *apply* their knowledge of how chemicals cycle through the living and nonliving components of an ecosystem. Specifically, this investigation gives students an opportunity to explore their natural surroundings where they live and gain an appreciation for the many factors involved in determining the health of the environment. This lab gives students an opportunity to understand how energy and matter flow through systems and how changes in an ecosystem can affect the stability of that system. Students will also have the opportunity to reflect on the difference between observations and inferences and between data and evidence.

The Content

Ecosystems include all the living and nonliving things in a certain area. All living things in an ecosystem are called *biotic factors*. Biotic factors include the plants and animals in the ecosystem, as well as smaller organisms such as bacteria and fungi. All nonliving things in an ecosystem are referred to as *abiotic factors*. Abiotic factors include the water, soil, rocks, and air found in the ecosystem, as well as chemicals. Water or air in an ecosystem is made up of many different chemicals. They include chemicals that are important for humans and animals to survive, such as the oxygen (O_2) we breathe and water molecules (H_2O). The air also contains other gases that other organisms use, such as carbon dioxide (CO_2) used by plants and nitrogen gas (N_2) used by bacteria. As living things use these chemicals for getting energy to survive, they will also release other chemicals that have similar elements, such as urea ($CO(NH_2)_2$) and phosphates (PO_4^{3-}). Living things release these chemicals into an ecosystem, usually through waste they produce or through decomposition after they die. All of this activity means that different chemicals move between living and nonliving parts of an ecosystem.

The patterns of movement of matter through the living and nonliving parts of an ecosystem are known as *biogeochemical cycles*. The "bio" aspects of these cycles involve living things, and the "geo" aspects of these cycles involve nonliving things. All aspects of these cycles involve chemicals that are different forms of matter. There are several major chemical elements that cycle through ecosystems, including carbon, nitrogen, phosphorus, and sulfur. Figure 12.1 provides a representation of the way carbon cycles through an ecosystem. The figures in the Lab 12 Reference Sheet describe the movement of nitrogen and phosphorus through ecosystems. Scientists have developed these models to make sense of the many complex interactions that occur regularly in ecosystems. The models also help scientists to focus on certain chemicals and processes as they evaluate the health and stability of an ecosystem. These cycles are based on a fundamental scientific law, the *law of conservation of matter*. This law tells us that as matter moves through an ecosystem, it will not be created or destroyed. The matter in an ecosystem simply changes forms as it is converted from one type of chemical to another.

FIGURE 12.1

The carbon cycle in ecosystems

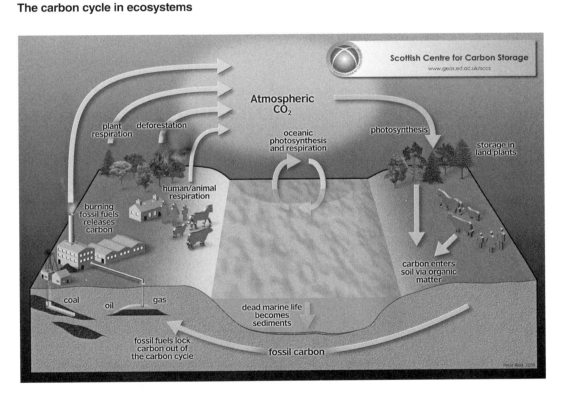

Humans also add chemicals into different ecosystems. In many cases, these chemicals are used to help humans get what they need from the ecosystem. An example of this is fertilizer. Farmers use fertilizer containing nitrogen- and phosphorus-based chemicals to help crops grow. Chemicals that are added to an ecosystem can benefit some organisms while also harming others. Humans also add chemicals into ecosystems in other ways, through processes such as the burning of fossil fuels and the treatment of wastewater. Although humans can benefit from these additions, they can also be harmed over longer periods of time as their activities cause an imbalance in an ecosystem. Examples of these events include instances of desertification where once-fertile land has changed to desert because overuse of the land for farming and grazing has killed off plant species and taken vital nutrients out of the soil. Eutrophication is another change in freshwater ecosystems that is partially due to excess fertilizer causing overgrowth of algae in those water systems. The algae consume significant amounts of O_2 dissolved in the water, leaving little for other aquatic plants and animals, and resulting in the death of those species.

Keeping natural cycles in balance helps keep ecosystems healthy. One way scientists figure out how healthy ecosystems are is through testing living and nonliving parts of them. It is a little more difficult to test living organisms in an ecosystem than it is to test nonliving parts of an ecosystem. Scientists will often test soil and water and other nonliving parts of an ecosystem, to understand how different chemicals are moving through an ecosystem. Using this information, scientists can determine if there is too much or too little of a certain chemical in an ecosystem. Looking at the levels of different chemicals helps them figure out how healthy the ecosystem is. The amounts of similar chemicals in different locations (water or soil) can show how matter is moving through an ecosystem. Testing the levels of chemicals in air, water, and soil is a major responsibility of government agencies charged with protecting the environment. These agencies have researched and determined ranges of chemical concentrations that are considered healthy.

Timeline

The instructional time needed to implement this lab investigation is 180–250 minutes. Appendix 2 (p. 355) provides options for implementing this lab investigation over several class periods. Option A (250 minutes) should be used if students are unfamiliar with scientific writing, because this option provides extra instructional time for scaffolding the writing process. You can scaffold the writing process by modeling, providing examples, and providing hints as students write each section of the report. Option B (180 minutes) should be used if students are familiar with scientific writing and have the skills needed to write an investigation report on their own. In option B, students complete stage 6 (writing the investigation report) and stage 8 (revising the investigation report) as homework.

Materials and Preparation

The materials needed to implement this investigation are listed in Table 12.1. You will need to decide whether to collect the soil and water samples yourself or have the students collect them. If you collect them, do so a few days before beginning the lab activity. Every effort should be made to make sure that samples are pesticide and/or herbicide free! When selecting locations to sample, try to find ecosystem locations that include both soil and water elements that connect. Ideal locations would be forests, farmlands, and riverbeds. The reasoning behind this is that the tests done on related water and soil can give a better sense of how the chemicals tested are moving through that system.

Soil and water quality test kits can be ordered from science supply companies. Kits should include tests for a variety of chemical characteristics such as pH, nitrates, phosphates, dissolved O_2, and turbidity. The test kits do not need to be professional grade for this investigation—a basic, inexpensive kit will suffice.

TABLE 12.1

Materials list

Item	Quantity
Consumables	
Soil sample from location A	1 per group
Soil sample from location B	1 per group
Soil sample from location C	1 per group
Water sample from location A	1 per group
Water sample from location B	1 per group
Water sample from location C	1 per group
Equipment and other materials	
Soil quality test kit (nitrogen, oxygenation, phosphorus, pH)	1 per group
Water quality test kit (pH, nitrates, phosphates, dissolved O_2, turbidity)	1 per group
Sanitized indirectly vented chemical-splash goggles	1 per student
Chemical-resistant apron	1 per student
Gloves	1 pair per student
Investigation Proposal C (optional)	1 per group
Whiteboard, 2' × 3'*	1 per group
Lab Handout	1 per student
Lab 12 Reference Sheet: The Nitrogen Cycle and the Phosphorus Cycle	1 per student
Peer-review guide	1 per student
Checkout Questions	1 per student

* As an alternative, students can use computer and presentation software such as Microsoft PowerPoint or Apple Keynote to create their arguments.

LAB 12

Safety Precautions

Follow all normal lab safety rules. In addition, take the following safety precautions:

1. Put on sanitized indirectly vented chemical-splash goggles and laboratory apron and gloves before starting the lab activity.

2. Handle all glassware with care to avoid breakage. Sharp glass edges can cut skin!

3. Review the important information on chemicals on the safety data sheet, and use caution when handling chemicals.

4. Wash hands with soap and water after completing the lab activity.

Topics for the Explicit and Reflective Discussion

Concepts That Can Be Used to Justify the Evidence

To provide an adequate justification of their evidence, students must explain why they included the evidence in their arguments and make the assumptions underlying their analysis and interpretation of the data explicit. In this investigation, students can use the following concepts to help justify their evidence:

- Ecosystems have both abiotic factors and biotic factors.
- Chemicals cycle through different components of an ecosystem but are never created nor destroyed; they just change forms (law of conservation of matter).
- Biogeochemical cycles can become unbalanced because of human activity.
- Chemicals may be helpful in soil but harmful in water.
- Unbalanced chemical cycles can permanently change ecosystems over long periods of time.

We recommend that you review these concepts during the explicit and reflective discussion to help students make this connection.

How to Design Better Investigations

It is important for students to reflect on the strengths and weaknesses of the investigation they designed during the explicit and reflective discussion. Students should therefore be encouraged to discuss ways to eliminate potential flaws, measurement errors, or sources of bias in their investigations. To help students be more reflective about the design of their investigation, you can ask the following questions:

- What were some of the strengths of your investigation? What made it scientific?
- What were some of the weaknesses of your investigation? What made it less scientific?

- If you were to do this investigation again, what would you do to address the weaknesses in your investigation? What could you do to make it more scientific?

Crosscutting Concepts

This investigation is aligned with two crosscutting concepts found in *A Framework for K–12 Science Education*, and you should review these concepts during the explicit and reflective discussion.

- *Energy and matter: Flows, cycles, and conservation:* Scientists must understand how energy and matter flow into, out of, and within a system in order to understand it and to understand how human activity can disrupt the natural balance of an ecosystem.
- *Stability and change:* It is critical to understand what makes a system stable or unstable and what controls rates of change in system.

The Nature of Science and the Nature of Scientific Inquiry

This investigation is aligned with two important concepts related to the *nature of science* (NOS) and the *nature of scientific inquiry* (NOSI), and you should review these concepts during the explicit and reflective discussion.

- *The difference between observations and inferences:* An observation is a descriptive statement about a natural phenomenon, whereas an inference is an interpretation of an observation. Students should also understand that current scientific knowledge and the perspectives of individual scientists guide both observations and inferences. Thus, different scientists can have different but equally valid interpretations of the same observations due to differences in their perspectives and background knowledge.
- *The difference between data and evidence in science:* Data are measurements, observations, and findings from other studies that are collected as part of an investigation. Evidence, in contrast, is analyzed data and an interpretation of the analysis.

Hints for Implementing the Lab

- Have the students review the chemicals tested for in each kit when they develop their method so they can determine which data they want to focus on and collect.
- If you provide the water and soil samples, also include some descriptions of the ecosystems you collected them from, including the presence and types of living things present, how close human activities and structures are located, and temperature of the soil and water when it was collected.

- If you have students collect the samples, have them record similar information or, at a minimum, have them take pictures of their sampling sites so they can describe some of this information later.

- Determining the "health" of an ecosystem involves many factors, so you may want to have a class discussion about the limits of ecosystem health that you can measure with the kits.

Topic Connections

Table 12.2 provides an overview of the scientific practices, crosscutting concepts, disciplinary core ideas, and supporting ideas at the heart of this lab investigation. In addition, it lists NOS and NOSI concepts for the explicit and reflective discussion. Finally, it lists literacy and mathematics skills (*CCSS ELA* and *CCSS Mathematics*) that are addressed during the investigation.

TABLE 12.2

Lab 12 alignment with standards

Scientific practices	• Asking questions and defining problems • Developing and using models • Planning and carrying out investigations • Analyzing and interpreting data • Using mathematics and computational thinking • Constructing explanations • Engaging in argument from evidence • Obtaining, evaluating, and communicating information
Crosscutting concepts	• Energy and matter: Flows, cycles, and conservation • Stability and change
Core idea	• LS2: Ecosystems: Interactions, energy, and dynamics
Supporting ideas	• Biotic and abiotic factors • Biogeochemical cycles • Law of conservation of matter • Chemical structures using the same element
NOS and NOSI concepts	• Observations and inferences • Difference between data and evidence
Literacy connections (CCSS ELA)	• *Reading:* Key ideas and details, craft and structure, integration of knowledge and ideas • *Writing:* Text types and purposes, production and distribution of writing, research to build and present knowledge, range of writing • *Speaking and listening:* Comprehension and collaboration, presentation of knowledge and ideas
Mathematics connections (CCSS Mathematics)	• Make sense of problems and persevere in solving them • Reason abstractly and quantitatively • Construct viable arguments and critique the reasoning of others • Model with mathematics • Use appropriate tools strategically

LAB 12

Lab Handout

Lab 12. Matter in Ecosystems: How Healthy Are Your Local Ecosystems?

Introduction

Ecosystems include all the living and nonliving things in a certain area. All living things in an ecosystem are called *biotic factors*. Biotic factors include the plants and animals in the ecosystem, as well as smaller organisms such as bacteria and fungi. All nonliving things in an ecosystem are referred to as *abiotic factors*. Abiotic factors include the water, soil, rocks, and air found in the ecosystem. Chemicals found in the ecosystem are also abiotic factors.

Water or air in an ecosystem is made up of many different chemicals. They include chemicals that are important for humans and animals to survive, such as the oxygen (O_2) we breathe and water molecules (H_2O). The air also contains other gases that other organisms use, such as carbon dioxide (CO_2) used by plants and nitrogen gas (N_2) used by bacteria. As living things use these chemicals for getting energy to survive, they will also release other chemicals that have similar elements, such as urea ($CO(NH_2)_2$) and phosphates (PO_4^{3-}). All of this activity means that different chemicals move between living and nonliving parts of an ecosystem.

The patterns of movement of matter through the living and nonliving parts of an ecosystem are known as *biogeochemical cycles*. The "bio" aspects of these cycles involve living things, and the "geo" aspects of these cycles involve nonliving things. All aspects of these cycles involve chemicals that are different forms of matter. The way scientists understand these cycles is based on a fundamental scientific law, the *law of conservation of matter*. This law tells us that as matter moves through an ecosystem, it will not be created or destroyed. The matter in an ecosystem simply changes forms as it is converted from one type of chemical to another.

Humans also add chemicals into different ecosystems. In many cases, these chemicals are used to help humans get what they need from the ecosystem. An example of this is fertilizer. Farmers use fertilizer containing nitrogen- and phosphorus-based chemicals to help crops grow. Chemicals that are added to an ecosystem can benefit some organisms while also harming others.

Keeping natural cycles in balance helps keep ecosystems healthy. One way scientists figure out how healthy ecosystems are is through testing living and nonliving parts of them. It is a little more difficult to test living organisms in an ecosystem than it is to test nonliving parts of an ecosystem. Scientists will often test soil and water and other nonliving parts of an ecosystem, to understand how different chemicals are moving through an ecosystem.

Using this information, scientists can determine if there is too much or too little of a certain chemical in an ecosystem. Looking at the levels of different chemicals helps them figure out how healthy the ecosystem is. The amounts of similar chemicals in different locations (water or soil) can show how matter is moving through an ecosystem.

Your Task

Conduct an investigation of ecosystems where you live and how healthy they are. Look at ecosystems that include both soil and water sources.

The guiding question of this investigation is, **How healthy are your local ecosystems?**

Materials

You may use any of the following materials during your investigation:

Consumables	Equipment
• Samples of soil from three locations in your area • Samples of water from three locations in your area	• Soil quality test kit (nitrogen, oxygenation, phosphorus, pH) • Water quality test kit (pH, nitrates, phosphates, dissolved O_2, turbidity) • Sanitized indirectly vented chemical-splash goggles • Chemical-resistant apron • Gloves • Lab 12 Reference Sheet

Safety Precautions

Follow all normal lab safety rules. In addition, take the following safety precautions:

1. Put on sanitized indirectly vented chemical-splash goggles and laboratory apron and gloves before starting the lab activity.

2. Handle all glassware with care to avoid breakage. Sharp glass edges can cut skin!

3. Review the important information on chemicals on the safety data sheet, and use caution when handling chemicals.

4. Wash hands with soap and water after completing the lab activity.

Investigation Proposal Required?   ☐ Yes   ☐ No

Getting Started

To answer the guiding question, you will need to analyze water and soil samples taken from local ecosystems and use background information about chemical cycling in ecosystems. To accomplish this task, you must first determine what type of data you need to

LAB 12

collect, how you will collect it, and how you will analyze it. To determine *what type of data you need to collect*, think about the following questions:

- What type of information do I need to collect from the Lab 12 Reference Sheet?
- What type of tests do I need to determine the quality of the water and soil samples? (*Hint:* Be sure to follow all directions given in the water and soil test kits.)
- What type of measurements or observations will you need to record during your investigation?

To determine *how you will collect your data*, think about the following questions:

- What will serve as a control (or comparison) condition?
- How will you make sure that your data are of high quality (i.e., how will you reduce error)?
- How will you keep track of the data you collect and how will you organize the data?

To determine *how you will analyze your data*, think about the following questions:

- What type of calculations will you need to make?
- What type of graph could you create to help make sense of your data?

Connections to Crosscutting Concepts, the Nature of Science, and the Nature of Scientific Inquiry

As you work through your investigation, be sure to think about

- how energy and matter flow into, out of, within, and through a system;
- how changes to different parts of ecosystems affect their stability;
- how observations and inferences are different but related to each other; and
- the difference between data collected in an investigation and evidence created in an investigation.

Initial Argument

Once your group has finished collecting and analyzing your data, you will need to develop an initial argument. Your argument must include a claim, evidence to support your claim, and a justification of the evidence. The claim is your group's answer to the guiding question. The evidence is an analysis and interpretation of your data. Finally, the justification of the evidence is why your group thinks the evidence matters. The justification of the evidence is important because scientists can use different kinds of evidence to support their claims. Your group will create your initial argument on a whiteboard. Your whiteboard should include all the information shown in Figure L12.1.

Argumentation Session

The argumentation session allows all of the groups to share their arguments. One member of each group will stay at the lab station to share that group's argument, while the other members of the group go to the other lab stations one at a time to listen to and critique the arguments developed by their classmates. This is similar to how scientists present their arguments to other scientists at conferences. If you are responsible for critiquing your classmates' arguments, your goal is to look for mistakes so these mistakes can be fixed and they can make their argument better. The argumentation session is also a good time to think about ways you can make your initial argument better. Scientists must share and critique arguments like this to develop new ideas.

FIGURE L12.1

Argument presentation on a whiteboard

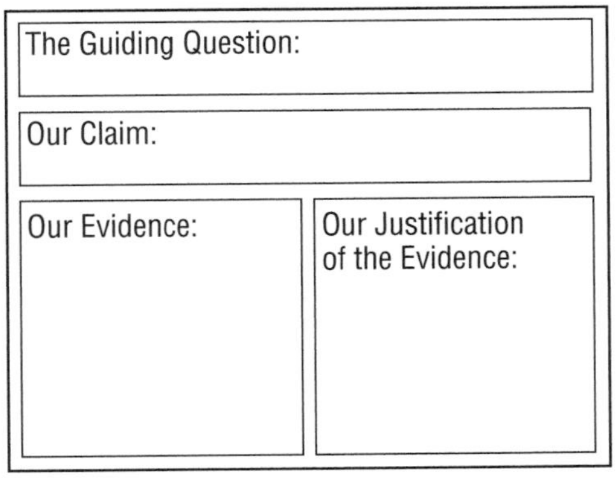

The Guiding Question:	
Our Claim:	
Our Evidence:	Our Justification of the Evidence:

To critique an argument, you might need more information than what is included on the whiteboard. You will therefore need to ask the presenter lots of questions. Here are some good questions to ask:

- What did your group do to collect the data? Why do you think that way is the best way to do it?
- What did your group do to analyze the data? Why did your group decide to analyze it that way?
- What other ways of analyzing and interpreting the data did your group talk about?
- What did your group do to make sure that these calculations are correct?
- Why did your group decide to present your evidence in that way?
- What other claims did your group discuss before you decided on that one? Why did your group abandon those other ideas?
- How sure are you that your group's claim is accurate? What could you do to be more certain?

Once the argumentation session is complete, you will have a chance to meet with your group and revise your original argument. Your group might need to gather more data or design a way to test one or more alternative claims as part of this process. Remember, your goal at this stage of the investigation is to develop the most valid or acceptable answer to the research question!

Report

Once you have completed your research, you will need to prepare an investigation report that consists of three sections that provide answers to the following questions:

LAB 12

1. What question were you trying to answer and why?

2. What did you do during your investigation and why did you conduct your investigation in this way?

3. What is your argument?

Your report should answer these questions in two pages or less. The report must be typed, and any diagrams, figures, or tables should be embedded into the document. Be sure to write in a persuasive style; you are trying to convince others that your claim is acceptable or valid!

National Science Teachers Association

Lab 12 Reference Sheet

The Nitrogen Cycle and the Phosphorus Cycle

Figure R12.1 shows the different forms that nitrogen gets combined into as it moves through an ecosystem. Nitrogen is a basic chemical needed for life. One of the most important functions of nitrogen is being one of the elements that form DNA. DNA is the molecule that allows living things to grow and reproduce. Nitrogen is also an important piece for building proteins. All living things in an ecosystem have nitrogen, and nitrogen moves from living to nonliving parts of an ecosystem in different forms and ways.

FIGURE R12.1

The nitrogen cycle

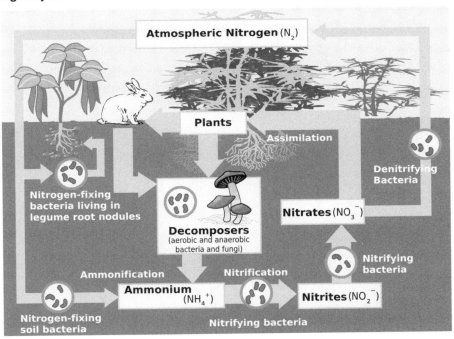

Plants and animals move nitrogen into air, water, and soil through the waste products they make. They also release nitrogen into an ecosystem when they die and their bodies decompose. Decomposing involves the breakdown of organic material into smaller chemicals. Different kinds of bacteria are very important in changing the forms of nitrogen that move through the nonliving parts of an ecosystem. Let's look at the different forms of nitrogen found in the nitrogen cycle.

LAB 12

Nitrogen gas (N_2): Nitrogen gas is the most common chemical in the air we breathe. However, plants and animals do not usually use nitrogen gas. Certain bacteria in soil can convert nitrogen gas into forms that plants and other bacteria can use. These bacteria are called *nitrogen-fixing.*

Ammonium (NH_4^+): Ammonium is a chemical you may have heard of in household cleaners. Ammonium moves into the soil when nitrogen-fixing bacteria change N_2 into NH_4^+. Ammonium is also produced when living things decompose. Ammonium in soil can be absorbed by plants through their roots and by other kinds of bacteria. Plants use ammonium as a source for nitrogen in their proteins and DNA. High levels of ammonium in water can poison fish and other living things there.

Nitrite (NO_2^-): Nitrite is a very unstable form for nitrogen in soil. Certain bacteria, called *nitrifying bacteria,* convert ammonium in the soil into nitrite. However, nitrite is usually changed into another form known as nitrate (see the next paragraph). This change is also done by nitrifying bacteria. Nitrite can be found in water sources. High levels of nitrite can cause diseases in some fish species.

Nitrate (NO_3^-): Nitrate is a more stable nitrogen molecule found in both soil and water. Nitrates can be absorbed by plants through their roots. Plants also use nitrates as a source for nitrogen in their proteins and DNA. Nitrates can be converted back into N_2 through another kind of bacteria called *denitrifying bacteria.* Nitrates are typically found in both soil and water. Nitrate levels in water are important to monitor. High levels of nitrates in drinking water can lead to serious health conditions in babies. High nitrate levels in water can also help algae grow out of control. Overgrowing algae can decrease the amount of nutrients, like oxygen, that other plants and animals use. As the algae continue to grow, the other plants and animals will die.

Figure R12.2 shows how phosphorus moves through an ecosystem, usually in the form of phosphate. Phosphate is an important molecule that living things use to build their DNA. Phosphorus is not typically found in the air in an ecosystem. Phosphate is mostly found in rocks and soil. Phosphate present in soil can be absorbed by plants to use. Animals eat the plants to get the phosphate they need. Rocks can be worn away by water rushing over them for long periods of time. As the water wears away the rock, phosphate is released into rivers, streams, and lakes. Animals and plants in the water can absorb the phosphate for their use. Phosphate that is not used by living things will fall to the bottom of the bodies of water as sediment. Over long periods of time, the phosphate sediment will change form into rocks again. Many fertilizers used by farmers contain high levels of phosphate. Phosphate helps algae grow out of control in fresh water. Overgrowing algae can decrease the amount of nutrients, like oxygen, that other plants and animals use. As the algae continue to grow, the other plants and animals will die.

FIGURE R12.2 _____

The phosphorus cycle

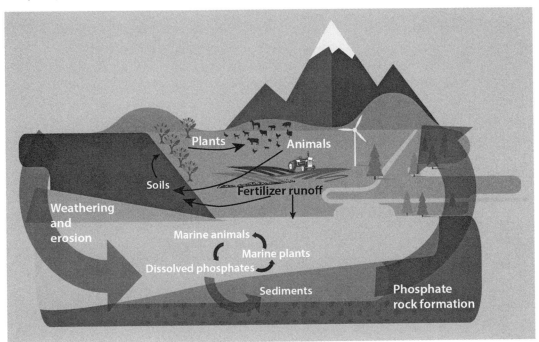

LAB 12

Lab 12. Matter in Ecosystems: How Healthy Are Your Local Ecosystems?

1. Farmer Johnson has two small ponds on his property. The first pond is next to a cornfield near the top of a hill. It has lots of fish and some vegetation. The second pond is also near the cornfield, but farther down the hill. The second pond has no fish and is almost full of vegetation. Use what you know about how matter moves through an ecosystem to explain why the two ponds have very different characteristics.

2. Organic farming involves growing fruits and vegetables without using any chemicals on the crops like pesticides (chemicals that kill bugs) or herbicides (chemicals that kill unwanted weeds). Use what you know about matter and how it moves through an ecosystem to explain why organic farming is a popular choice.

3. Observations are more important in science than inferences.

   a. I agree with this statement.

   b. I disagree with this statement.

   Explain your answer, using an example from your investigation about matter in ecosystems.

4. In science, data and evidence are the same thing.

   a. I agree with this statement.

   b. I disagree with this statement.

   Explain your answer, using an example from your investigation about matter in ecosystems.

5. When scientists study the natural world, they often need to keep track of how matter and energy moves through a system. Explain why understanding the flow of matter and energy is important, using an example from your investigation about matter in ecosystems.

6. Scientists study complex systems that have many related parts. Change in one aspect of a system can have impacts on many other parts of the system. Explain how changes in one part of an ecosystem can influence other parts of the system, using an example from your investigation about matter in ecosystems.

LAB 13

Lab 13. Carbon Cycling: Which Carbon Cycle Process Affects Atmospheric Carbon the Most?

Purpose

The purpose of this lab is for students to *apply* their knowledge of how chemicals, specifically carbon, cycle through the living and nonliving components of an ecosystem. Specifically, this investigation gives students an opportunity to explore how two major processes in carbon cycling, photosynthesis and combustion, influence the amount of atmospheric carbon. Teachers will need to help students understand the use of computer simulations to forecast environmental changes. This lab gives students an opportunity to understand how scientific phenomena occur over different scales of time and how changes in an ecosystem can affect the stability of that system. Students will also have the opportunity to reflect on the difference between data and evidence and how scientists use different methods in their investigations, including the use of computer simulations.

The Content

Nutrients are chemicals that are essential for plant and animal growth. Animals get nutrients by eating plants or other animals. Plants get nutrients from the nonliving parts of an ecosystem, such as water and soil. Nutrients often cycle through an ecosystem, which means they change forms and move through both living and nonliving parts. Carbon, nitrogen, and phosphorus are chemicals that cycle through an ecosystem. The carbon cycle is one of the most important cycles that support life in an ecosystem. Figure 13.1 illustrates the sources and storage of carbon as it cycles through an ecosystem.

The following list traces the path of a carbon atom:

- Atmospheric carbon dioxide gas (CO_2) is a source of carbon in the cycle. The atmosphere is the layer of gases that surround the planet. CO_2 passes into the ecosystem through living organisms and nonliving elements through different processes.
- CO_2 leaves the atmosphere through photosynthesis and decomposition.
  - *Photosynthesis:* This chemical process is used by plants to make food (in the form of sugar), using CO_2 from the atmosphere. The carbon material is eventually moved into other organisms through the food web. Forests and other areas with a lot of plant life are where the greatest amounts of photosynthesis happen.

FIGURE 13.1

The carbon cycle in ecosystems

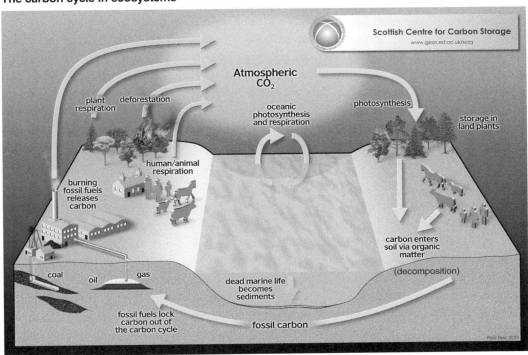

- *Decomposition:* The carbon stored in the bodies of living things, both plants and animals, will continue to be stored in soil after they die. Their bodies will break down into simpler carbon molecules, referred to as *organic material.* This organic material can remain in the soil for long periods of time, ultimately making *fossil fuels*—carbon-rich materials made from the broken-down remains of plants and animals that lived millions of years ago. These fuels include coal, oil, and natural gas. Humans use these fuels to provide energy for technology and communities.

- CO_2 is released into the atmosphere through weathering, respiration, and combustion.

  - *Weathering:* Calcium carbonate ($CaCO_3$) is in rocks. When rocks combine with water, carbonic acid (H_2CO_3) is formed and it then splits into carbonate, bicarbonate, and hydrogen ions.

  - *Respiration:* Sugar + oxygen gas (O_2) will produce $CO_2 + H_2O$. This is the process most living things use to produce the energy they need to survive. Living things on land and in water use this process, releasing CO_2 into both the air and bodies of water.

• *Combustion:* When trees are burned, CO_2 is released. The burning of large areas of trees is known as *deforestation.* Human activities add more CO_2 into the atmosphere through activities like the burning of fossil fuels.

There can be more CO_2 in the atmosphere than plants can use in photosynthesis. This extra carbon in the atmosphere can lead to several changes in the environment. Extra CO_2 can trap heat energy coming from the planet's surface. As this heat energy gets trapped, it can raise the global temperature. This heating due to extra CO_2 is known as the *greenhouse effect.* Scientists have demonstrated that the rise in global temperature leads to changes in the climate of different regions of the world. *Climate* includes trends in different conditions in the environment that happen over long periods of time; these trends include average temperature changes and amounts of rain, humidity, and air pressure.

Several major international bodies have recognized the relationship between increasing levels of atmospheric CO_2 and changes in climate patterns across the globe. The most well-known and authoritative group focused on these issues is the Intergovernmental Panel on Climate Change (IPCC). This group has reviewed and analyzed large amounts of data to reach several important conclusions regarding the nature and causes of climate change. Although some of their findings are disputed in political and economic discussions, a large majority of the scientific community is in strong agreement with the findings. This group continues to analyze and report on findings related to climate change and to make recommendations for actions to mediate the effects of climate change. Recently, the IPCC also recognized another atmospheric gas that was contributing to climate change. Methane (CH_4) gas traps even more heat in the atmosphere because of its chemical structure. Sources of methane gas include the natural gas reserves that are accessed through human activities, as well as the gas naturally produced by animals, like cows, through their digestion of food material.

Extra atmospheric CO_2 gas can also dissolve into bodies of water, especially oceans. When CO_2 combines with water, H_2CO_3 is formed. Over time, this process leads to lower pH levels in the ocean, which is called *ocean acidification.* Lower pH in oceans can hurt the living things in the water and decrease the amount of O_2 available. This effect of rising atmospheric CO_2 levels is often not discussed as much as atmospheric effects because of the longer time frame for the phenomenon. Also, ocean acidification is not solely attributable to atmospheric CO_2. Other chemicals that are released into the oceans through the dumping of wastewater and petroleum leakage from ships and tankers contribute to acidification. This acidification has been connected to the continued killing of large areas of the great coral reefs in several locations across the globe, including those found in the Florida Keys in the United States and the Great Barrier Reef in Australia.

Timeline

The instructional time needed to implement this lab investigation is 180–250 minutes. Appendix 2 (p. 355) provides options for implementing this lab investigation over several class periods. Option A (250 minutes) should be used if students are unfamiliar with scientific writing, because this option provides extra instructional time for scaffolding the writing process. You can scaffold the writing process by modeling, providing examples, and providing hints as students write each section of the report. Option B (180 minutes) should be used if students are familiar with scientific writing and have the skills needed to write an investigation report on their own. In option B, students complete stage 6 (writing the investigation report) and stage 8 (revising the investigation report) as homework.

Materials and Preparation

The materials needed to implement this investigation are listed in Table 13.1. The *Carbon Lab* simulation from Annenberg Learner, available at *www.learner.org/courses/envsci/interactives/carbon/carbon.html*, is free to use and can be run online using an internet browser. You can use a computer lab or a classroom set of laptops. *Note:* The simulation used in this activity does not operate as fluidly on tablets.

You should access the website and learn how the simulation works before beginning the lab investigation. In addition, it is important to check if students can access and use the simulation from a school computer, because some schools have set up firewalls and other restrictions on web browsing.

TABLE 13.1

Materials list

Item	Quantity
Computer with internet access	1 per group
Investigation Proposal A (optional)	1 per group
Whiteboard, 2' × 3'*	1 per group
Lab Handout	1 per student
Peer-review guide	1 per student
Checkout Questions	1 per student

* As an alternative, students can use computer and presentation software such as Microsoft PowerPoint or Apple Keynote to create their arguments.

Safety Precautions

Follow all normal lab safety rules.

LAB 13

Topics for the Explicit and Reflective Discussion

Concepts That Can Be Used to Justify the Evidence

To provide an adequate justification of their evidence, students must explain why they included the evidence in their arguments and make the assumptions underlying their analysis and interpretation of the data explicit. In this investigation, students can use the following concepts to help justify their evidence:

- Chemicals cycle through different components of an ecosystem but are never created nor destroyed; they just change forms (this is the law of conservation of matter).
- Carbon cycles through ecosystems by two major processes: photosynthesis and combustion.
- With deforestation, photosynthesis in the environment is decreased while combustion is increased.
- Combustion releases CO_2 gas from fossil fuels and living things, such as trees.

We recommend that you review these concepts during the explicit and reflective discussion to help students make this connection.

How to Design Better Investigations

It is important for students to reflect on the strengths and weaknesses of the investigation they designed during the explicit and reflective discussion. Students should therefore be encouraged to discuss ways to eliminate potential flaws, measurement errors, or sources of bias in their investigations. To help students be more reflective about the design of their investigation, you can ask the following questions:

- What were some of the strengths of your investigation? What made it scientific?
- What were some of the weaknesses of your investigation? What made it less scientific?
- If you were to do this investigation again, what would you do to address the weaknesses in your investigation? What could you do to make it more scientific?

Crosscutting Concepts

This investigation is aligned with two crosscutting concepts found in *A Framework for K–12 Science Education,* and you should review these concepts during the explicit and reflective discussion.

- *Scale, proportion, and quantity:* It is critical for scientists to be able to recognize what is relevant at different sizes, time frames, and scales. Scientists must also be able to recognize proportional relationships between categories or quantities.

- *Stability and change:* It is critical for scientists to understand what makes a system stable or unstable and how different processes affect rates of change in system.

The Nature of Science and the Nature of Scientific Inquiry

This investigation is aligned with two important concepts related to the *nature of science* (NOS) and the *nature of scientific inquiry* (NOSI), and you should review these concepts during the explicit and reflective discussion.

- *The difference between data and evidence in science:* Data are measurements, observations, and findings from other studies that are collected as part of an investigation. Evidence, in contrast, is analyzed data and an interpretation of the analysis.

- *Methods used in scientific investigations:* Examples of methods include experiments, systematic observations of a phenomenon, literature reviews, and analysis of existing data sets; the choice of method depends on the objectives of the research. There is no universal step-by-step scientific method that all scientists follow; rather, different scientific disciplines (e.g., chemistry vs. physics) and fields within a discipline (e.g., ecology vs. molecular biology) use different types of methods, use different core theories, and rely on different standards to develop scientific knowledge.

Hints for Implementing the Lab

- Consider doing a class demonstration of the website during stage 1 so you can introduce students to the sliding scales and have them discuss what each scale means in relation to the processes described in the Lab Handout.

- As groups develop their methods, draw their attention to what variable they will use as their outcome in their investigation. They can use the graph that shows atmospheric CO_2 levels over the years. They could also use the graphics showing the gigaton (GT) amounts of carbon in different parts of the environment.

- Consider providing extra information on humans' use of fossil fuels, especially if you have not discussed these topics to any extent before this investigation.

- Remind students that computer simulations such as the one used in this activity are constructed based on data collected through previous investigations. You may want to show them some of the Intergovernmental Panel on Climate Change reports, which also condense a lot of the studies used to build simulations like these.

Topic Connections

Table 13.2 (p. 218) provides an overview of the scientific practices, crosscutting concepts, disciplinary core ideas, and supporting ideas at the heart of this lab investigation. In addition, it lists NOS and NOSI concepts for the explicit and reflective discussion. Finally, it

LAB 13

lists literacy and mathematics skills (*CCSS ELA* and *CCSS Mathematics*) that are addressed during the investigation.

TABLE 13.2

Lab 13 alignment with standards

Scientific practices	• Asking questions and defining problems • Developing and using models • Planning and carrying out investigations • Analyzing and interpreting data • Using mathematics and computational thinking • Constructing explanations • Engaging in argument from evidence • Obtaining, evaluating, and communicating information
Crosscutting concepts	• Scale, proportion, and quantity • Stability and change
Core idea	• LS2: Ecosystems: Interactions, energy, and dynamics
Supporting ideas	• Carbon cycling • Photosynthesis • Combustion • Law of conservation of matter • Human activities • Deforestation
NOS and NOSI concepts	• Difference between data and evidence • Methods used in scientific investigations
Literacy connections (*CCSS ELA*)	• *Reading:* Key ideas and details, craft and structure, integration of knowledge and ideas • *Writing:* Text types and purposes, production and distribution of writing, research to build and present knowledge, range of writing • *Speaking and listening:* Comprehension and collaboration, presentation of knowledge and ideas
Mathematics connections (*CCSS Mathematics*)	• Make sense of problems and persevere in solving them • Reason abstractly and quantitatively • Construct viable arguments and critique the reasoning of others • Model with mathematics • Use appropriate tools strategically • Look for and express regularity in repeated reasoning

Reference

Annenberg Learner. 2015. Carbon Lab. Indianapolis, IN: Annenberg Learner. *www.learner.org/courses/envsci/interactives/carbon/carbon.html.*

Lab Handout

Lab 13. Carbon Cycling: Which Carbon Cycle Process Affects Atmospheric Carbon the Most?

Introduction

Organisms live together in ecosystems and rely on each other for food. All living things also require different amounts and kinds of *nutrients*, including nonliving factors in an ecosystem. Nutrients are chemicals that are essential for plant and animal growth. Animals get nutrients by eating plants or other animals. Plants get nutrients from the nonliving parts of an ecosystem, such as water and soil. In fact, nutrients often cycle through an ecosystem, which means they change forms and move through both living and nonliving parts. Carbon, nitrogen, and phosphorus are chemicals that cycle through an ecosystem. The carbon cycle is one of the most important cycles that support life in an ecosystem. Figure L13.1 illustrates the sources and storage of carbon as it cycles through an ecosystem.

FIGURE L13.1 _____

The carbon cycle in ecosystems

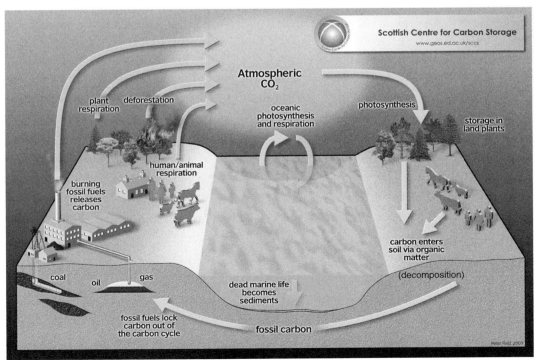

LAB 13

The following list traces the path of a carbon atom:

- Atmospheric carbon dioxide gas (CO_2) is a source of carbon in the cycle. The atmosphere is the layer of gases that surround the planet. CO_2 passes into the ecosystem through living organisms and nonliving elements through different processes.

- CO_2 leaves the atmosphere through *photosynthesis*, which is the chemical process used by plants to make food (in the form of sugar) using CO_2 from the atmosphere. This carbon material is eventually moved into other organisms through the food web. Forests and other areas with a lot of plant life are where the greatest amounts of photosynthesis happen.

- CO_2 can be released into the atmosphere through *combustion*. When trees are burned, CO_2 is released. The burning of large areas of trees is known as *deforestation*. Human activities add more CO_2 into the atmosphere through activities like the burning of *fossil fuels*, which are carbon-rich materials made from the broken-down remains of plants and animals that lived millions of years ago. These fuels include coal, oil, and natural gas. Humans use these fuels to provide energy for technology and communities.

There can be more CO_2 in the atmosphere than plants can use in photosynthesis. This extra carbon in the atmosphere can lead to several changes in the environment. Extra CO_2 can trap heat energy coming from the planet's surface. As this heat energy gets trapped, it can raise the global temperature. This heating due to extra CO_2 is known as the *greenhouse effect*. Scientists have demonstrated that the rise in global temperature leads to changes in the climate of different regions of the world. *Climate* includes trends in different conditions in the environment that happen over long periods of time. Extra atmospheric CO_2 can also dissolve into bodies of water, especially oceans. When CO_2 combines with water, carbonic acid (H_2CO_3) is formed. Over time, this process leads to lower pH levels in the ocean, which is called *ocean acidification*.

Your Task

Use a computer simulation to determine whether atmospheric CO_2 will decrease more from burning fossil fuels or from changing the amount of deforestation.

The guiding question of this investigation is, **Which carbon cycle process affects atmospheric carbon the most?**

Materials

You will use an online simulation called *Carbon Lab* to conduct your investigation. You can find the simulation by going to the following website: *www.learner.org/courses/envsci/ interactives/carbon/carbon.html*.

Safety Precautions

Follow all normal lab safety rules.

Investigation Proposal Required?   ☐ Yes   ☐ No

Getting Started

Use the *Carbon Lab* simulation to estimate how decreasing plant life through deforestation and increasing the fossil fuel use will affect other parts of the planet. You will need to run simulations using different settings for increasing and decreasing the amount of fossil fuels burned. You will also need to run simulations using different settings for increasing and decreasing the amount of deforestation happening. You can adjust these settings using the sliding scales on the screen. Once you have set the scales, then you need to press the "Run Decade" button. One decade will show what happens to atmospheric CO_2 after 10 years at the settings you choose. The image will also show how CO_2 is distributed in different parts of the environment after that decade as well. You will need to run many decades to see what happens over longer periods of time. You may notice that the units of CO_2 provided are GT, which stands for gigatons; 1 GT is equal to 1,000,000,000,000,000 grams of a substance.

To answer the guiding question, you will need to observe changes in the amount of atmospheric carbon. You can also observe how much carbon is stored in different parts of the environment. To accomplish this task, you must first determine what type of data you need to collect, how you will collect it, and how you will analyze it. To determine *what type of data you will need to collect*, think about the following questions:

- What type of data is available in the simulation?
- What information about carbon cycle processes is provided by different data sources?
- What type of measurements or observations will you need to record during your investigation?

To determine *how you will collect your data*, think about the following questions:

- What will serve as a control (or comparison) condition?
- How will you make sure that your data are of high quality (i.e., how will you reduce error)?
- How will you keep track of the data you collect and how will you organize the data?

To determine *how you will analyze your data*, think about the following questions:

- What type of calculations will you need to make?
- What type of graph could you create to help make sense of your data?

LAB 13

Connections to Crosscutting Concepts, the Nature of Science, and the Nature of Scientific Inquiry

As you work through your investigation, be sure to think about

- how events of scientific interest occur over different scales of time,
- how changes to different parts of ecosystems affect their stability,
- the difference between data collected in an investigation and evidence created in an investigation, and
- how scientists use different methods in their investigations.

Initial Argument

Once your group has finished collecting and analyzing your data, you will need to develop an initial argument. Your argument must include a claim, evidence to support your claim, and a justification of the evidence. The claim is your group's answer to the guiding question. The evidence is an analysis and interpretation of your data. Finally, the justification of the evidence is why your group thinks the evidence matters. The justification of the evidence is important because scientists can use different kinds of evidence to support their claims. Your group will create your initial argument on a whiteboard. Your whiteboard should include all the information shown in Figure L13.2.

FIGURE L13.2

Argument presentation on a whiteboard

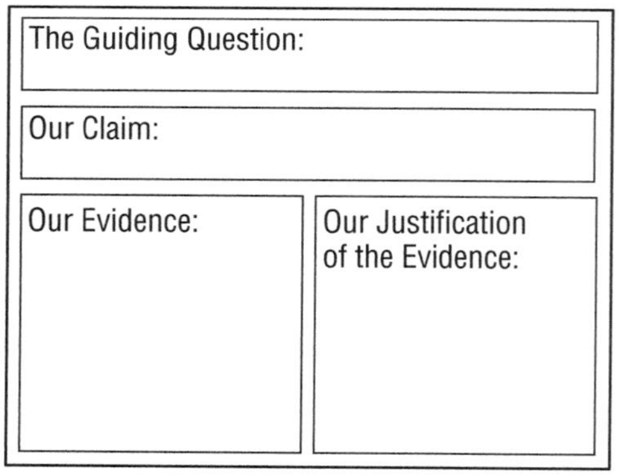

The Guiding Question:	
Our Claim:	
Our Evidence:	Our Justification of the Evidence:

Argumentation Session

The argumentation session allows all of the groups to share their arguments. One member of each group will stay at the lab station to share that group's argument, while the other members of the group go to the other lab stations one at a time to listen to and critique the arguments developed by their classmates. This is similar to how scientists present their arguments to other scientists at conferences. If you are responsible for critiquing your classmates' arguments, your goal is to look for mistakes so these mistakes can be fixed and they can make their argument better. The argumentation session is also a good time to think about ways you can make your initial argument better. Scientists must share and critique arguments like this to develop new ideas.

To critique an argument, you might need more information than what is included on the whiteboard. You will therefore need to ask the presenter lots of questions. Here are some good questions to ask:

- What did your group do to collect the data? Why do you think that way is the best way to do it?

- What did your group do to analyze the data? Why did your group decide to analyze it that way?
- What other ways of analyzing and interpreting the data did your group talk about?
- What did your group do to make sure that these calculations are correct?
- Why did your group decide to present your evidence in that way?
- What other claims did your group discuss before you decided on that one? Why did your group abandon those other ideas?
- How sure are you that your group's claim is accurate? What could you do to be more certain?

Once the argumentation session is complete, you will have a chance to meet with your group and revise your original argument. Your group might need to gather more data or design a way to test one or more alternative claims as part of this process. Remember, your goal at this stage of the investigation is to develop the most valid or acceptable answer to the research question!

Report

Once you have completed your research, you will need to prepare an investigation report that consists of three sections that provide answers to the following questions:

1. What question were you trying to answer and why?
2. What did you do during your investigation and why did you conduct your investigation in this way?
3. What is your argument?

Your report should answer these questions in two pages or less. The report must be typed, and any diagrams, figures, or tables should be embedded into the document. Be sure to write in a persuasive style; you are trying to convince others that your claim is acceptable or valid!

LAB 13

Checkout Questions

Lab 13. Carbon Cycling: Which Carbon Cycle Process Affects Atmospheric Carbon the Most?

1. Zachary thinks that deforestation only affects the levels of carbon dioxide gas (CO_2) in the atmosphere because of burning the trees that get cut down. Elizabeth thinks cutting the trees down is enough to affect the levels of CO_2 in the atmosphere. Using what you know about the carbon cycle, provide supporting evidence for either Zachary or Elizabeth.

2. Some students in science class were discussing photosynthesis and the carbon cycle. The students were confused about the impact of burning fossil fuels. Burning fossil fuels releases CO_2 into the atmosphere, and plants use CO_2 during photosynthesis to grow. The students concluded that burning fossil fuels is a good thing because all the plants and trees will grow bigger. Use what you know about the carbon cycle and atmospheric CO_2 to help clear up their confusion.

3. In science, data and evidence are the same thing.

    a. I agree with this statement.

    b. I disagree with this statement.

    Explain your answer, using an example from your investigation about carbon cycling.

4. When scientists study complex systems like the carbon cycle, they must focus on processes that occur on very different time scales. For example, trees can take 100 years to reach full size, fossil fuels have formed over millions of years, photosynthesis is an ongoing process, and trees can be cut down and burned in just a few seconds or minutes. Explain why understanding the different time scales for events is important, using an example from your investigation about carbon cycling.

5. Scientists study complex systems that have many related parts. Change in one aspect of a system can have impacts on many other parts of the system. Explain how changes in one aspect of the carbon cycle influence other aspects of the system.

SECTION 4

Life Sciences
Core Idea 3

Heredity: Inheritance and Variation in Traits

Introduction Labs

LAB 14

Teacher Notes

Lab 14. Variation in Traits: How Do Beetle Traits Vary Within and Across Species?

Purpose

The purpose of this lab is to *introduce* students to the variation that naturally exists within a species and among different species that are closely related. Specifically, this investigation gives students an opportunity to explore the ways in which beetles vary in the traits they possess, mainly physically, but also related to their environment. Teachers will need to help students identify different traits, both physical and behavioral, that they can use to compare the different species focused on in this activity. This lab gives students an opportunity to learn how scientists focus on identifying patterns to make sense of the natural world and to learn about the connections between organisms' structures in their body and the functions they perform. Students will also have the opportunity to reflect on the difference between observations and inferences and how scientific knowledge develops over time as evidence continues to be collected to support or refute ideas.

The Content

Organisms differ from one another in several ways. When those differences are so great that organisms are unable to mate and produce fertile offspring, they are said to be of different *species*. Differences between species, or so-called *interspecies differences*, can be great or small. Some differences can be easily seen, such as different shapes of arms or legs. Others are not easy to observe, such as differences in the kinds of genes each species has. Differences also exist within a species; in other words, not all members of the same species are the same (e.g., not all gray whales are the same). These differences and similarities, both within and across species, are known as *variation*. Variation, both within and across species, is very important in creating the diversity of life that exists on the planet. Charles Darwin identified variation in organisms as a fundamental requirement that drives the process of natural selection.

Variation provides the raw material for evolutionary changes. Depending on time and environment, small variations in traits such as color and size can create a reproductive advantage, allowing a specific form of a species to survive better in that time and place. Those who survive better live long enough to reproduce and pass on those advantages to their offspring. Over several generations, those species with the advantageous trait may survive better than fellow members of their species and that trait becomes the most prominent in certain populations. As time continues, variation can allow for many new advantageous traits to emerge in this way, so much that organisms with enough different traits become reproductively isolated from similar organisms. When this type of branching results from natural selection acting on variation, new species are formed and the diversity of life expands; this process is called *speciation*. The ancestor species may also continue to exist, rather than be replaced by the new species, depending on the intensity of the competition between the two for limited resources present in their local environments. Further, while the new species becomes the dominant form in one environment, the ancestor species may remain the dominant form in other locations. Thus, variation is the fuel for evolution, but the small-scale processes (natural selection) and the large-scale ones (speciation) are influenced by the environments in which organisms live and the survival possible in those places. Observations about the differences and similarities between organisms, especially species, served as the foundation for the evidence Darwin developed to support his theories.

Even before Darwin developed his revolutionary ideas about evolution, scientists intently worked to make sense of and organize the vast diversity of living things they encountered in the world. To understand the variation present in living things, scientists developed a system of grouping organisms together based on their differences and similarities; this system is called *classification*. The system begins with a few large groups of organisms that share a few similar traits on the first level. The next level of classification splits those large groups into smaller groups using differences and similarities among the organisms in them. This second set of groups is then split up even further based on the variation among the organisms. Current classification systems use eight major levels to classify all the living things, as described in the next paragraph. Some of these levels also have sublevels, which further improves classification.

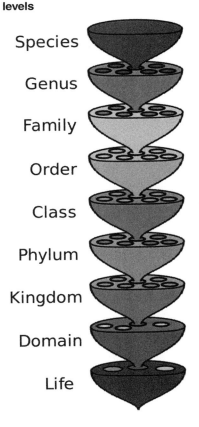

FIGURE 14.1

Current taxonomic classification levels

Species

Genus

Family

Order

Class

Phylum

Kingdom

Domain

Life

Taxonomy is the branch of science that deals with classifying organisms. The most prominent taxonomic classification system still in use was first developed by Carl Linnaeus in the early 1700s. His original system used six primary levels to group organisms into different classifications. However, as scientists have continued to discover more and more species, the system has been modified to reflect new relationships between organisms and groupings. Further necessitating this expansion was the development of new ways of comparing organisms beyond physical characteristics, especially through the more recent understanding of organisms' internal genetic structures. Current taxonomic classification schemes include eight different levels, beginning with domain, then kingdom, phylum, class, order, family, genus, and species (Figure 14.1). Several of these major levels are also divided into sublevels to further organize the diversity of life based on differences and similarities in their variations.

LAB 14

This lab investigation focuses on the beetle as a demonstration of variation within and across species. The name Beetle is given to an order of insects that all have elytra. The elytra are a hardened, sheathlike set of front wings, which usually cover the entire abdomen when the insect is not in flight. Beetles vary greatly in size. The largest is the Eastern Hercules beetle (*Dynastes tityus*), which grows to 16 cm (up to 6.3 inches) in length (Figure 14.2). Different species can have very different appearances; compare, for example, the Eastern Hercules beetle in Figure 14.2 with the European violet ground beetle in Figure 14.3. Other species may be less than 0.1 cm (less than 0.04 inch) long.

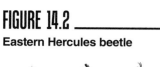

FIGURE 14.2 _____

Eastern Hercules beetle

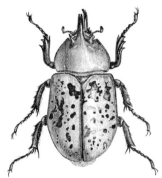

FIGURE 14.3_____

European violet ground beetle

The Beetle order is split into a number of smaller groups, which involve more similarities in shape between the group members. These smaller groups follow a strict hierarchy. The major levels and sublevels smaller than order are called suborder, family, subfamily, and genus. A genus is the smallest group of importance in the naming of individual species, although in some classifications generic groups may be further split into subgenera. The scientific name of a species includes first its genus and second its specific name. For example, the European violet ground beetle is called *Carabus violaceus*, meaning the species *violaceus* in the genus *Carabus.* (Genus and species names are always printed in italics. The genus name is capitalized, and the species name is lowercase [starts with small letter].) The full classification of the European violet ground beetle is as shown in Table 14.1.

The Beetle order embraces more species than any other group in the animal kingdom. At least 250,000 species are known. This represents more than one-quarter of all animal species and is far more than all mammal species combined. With all these different species, a lot of variation exists among beetles. As such, they serve as a great example for students to investigate naturally occurring variation among species.

National Science Teachers Association

TABLE 14.1

Classification of the European violet ground beetle

Classification level	Name	Description
Phylum	Arthropoda	Arthropod
Class	Insecta	Insect
Subclass	Pterygota	Winged insect
Order	Coleoptera	Beetle
Suborder	Adephaga	Carnivorous beetle
Family	Carabidae	Ground beetle
Genus	*Carabus*	*
Species	*violaceus*	Violet ground beetle

*There is no well-defined descriptor for beetle genera. The descriptor often refers to geographic location, but this specific genus is quite diverse and not restricted to only one region.

Timeline

The instructional time needed to implement this lab investigation is 130–200 minutes. Appendix 2 (p. 355) provides options for implementing this lab investigation over several class periods. Option C (200 minutes) should be used if students are unfamiliar with scientific writing, because this option provides extra instructional time for scaffolding the writing process. You can scaffold the writing process by modeling, providing examples, and providing hints as students write each section of the report. Option D (130 minutes) should be used if students are familiar with scientific writing and have the skills needed to write an investigation report on their own. In option D, students complete stage 6 (writing the investigation report) and stage 8 (revising the investigation report) as homework.

Materials and Preparation

The materials needed to implement this investigation are listed in Table 14.2 (p. 234). If you want to use more information sources than the Lab 14 Reference Sheet, you will need computers that can access the internet. You can use a computer lab or a classroom set of laptops. Two suggested websites are BugGuide (*www.bugguide.net*) and Wikispecies (*http://species.wikimedia.org*), but feel free to locate other websites that you think would be useful for your students during their investigation. Another option can be to order preserved samples of different beetle species from a science supply company such as Carolina or Ward's Science.

LAB 14

TABLE 14.2
Materials list

Item	Quantity
Preserved beetle samples of three species (optional)	2 per group
Computer with internet access	1 per group
Investigation Proposal A (optional)	1 per group
Whiteboard, 2' × 3'*	1 per group
Lab Handout	1 per student
Lab 14 Reference Sheet: Three Types of Beetles	2 per group
Peer-review guide	1 per student
Checkout Questions	1 per student

* As an alternative, students can use computer and presentation software such as Microsoft PowerPoint or Apple Keynote to create their arguments.

Safety Precautions

Follow all normal lab safety rules.

Topics for the Explicit and Reflective Discussion

Concepts That Can Be Used to Justify the Evidence

To provide an adequate justification of their evidence, students must explain why they included the evidence in their arguments and make the assumptions underlying their analysis and interpretation of the data explicit. In this investigation, students can use the following concepts to help justify their evidence:

- Variation in traits is used to classify organisms.

- Some physical traits can be unique to an individual. Other physical traits, like certain body structures, are important in distinguishing between species.

- Genetic traits are not easy to see, but they help explain how species are different. They can lead to different patterns of behaviors that can result in separate species.

- Taxonomy, including species categories, helps organize the multiple forms of similar life present on the planet.

We recommend that you review these concepts during the explicit and reflective discussion to help students make this connection.

How to Design Better Investigations

It is important for students to reflect on the strengths and weaknesses of the investigation they designed during the explicit and reflective discussion. Students should therefore be encouraged to discuss ways to eliminate potential flaws, measurement errors, or sources of bias in their investigations. To help students be more reflective about the design of their investigation, you can ask the following questions:

- What were some of the strengths of your investigation? What made it scientific?
- What were some of the weaknesses of your investigation? What made it less scientific?
- If you were to do this investigation again, what would you do to address the weaknesses in your investigation? What could you do to make it more scientific?

Crosscutting Concepts

This investigation is aligned with two crosscutting concepts found in *A Framework for K–12 Science Education,* and you should review these concepts during the explicit and reflective discussion.

- *Patterns:* Observed patterns in nature, such as similar structures with similar functions in organisms, guide the way scientists organize and classify life on Earth. Scientists also explore the relationships between and the underlying causes of the patterns they observe in nature.
- *Structure and function:* In nature, the way a living thing is structured or shaped determines how it functions and places limits on what it can and cannot do.

The Nature of Science and the Nature of Scientific Inquiry

This investigation is aligned with two important concepts related to the *nature of science* (NOS) and the *nature of scientific inquiry* (NOSI), and you should review these concepts during the explicit and reflective discussion.

- *The difference between observations and inferences:* An observation is a descriptive statement about a natural phenomenon, whereas an inference is an interpretation of an observation. Students should also understand that current scientific knowledge and the perspectives of individual scientists guide both observations and inferences. Thus, different scientists can have different but equally valid interpretations of the same observations due to differences in their perspectives and background knowledge.
- *Changes in scientific knowledge over time:* A person can have confidence in the validity of scientific knowledge but must also accept that scientific knowledge may be abandoned or modified in light of new evidence or because existing evidence has been reconceptualized by scientists. There are many examples in the history of

LAB 14

science of both evolutionary changes (i.e., the slow or gradual refinement of ideas) and revolutionary changes (i.e., the rapid abandonment of a well-established idea) in scientific knowledge.

Hints for Implementing the Lab

- Consider using other information sources to help students make comparisons among the beetle species, including informative websites.
- As groups make their comparisons, suggest that they look at differences in the shape of certain structures that are common across species.
- Remind students to think about whether individual variations within a species could help those individuals survive better than others.
- Remind students that creating charts or graphics can help to organize their observations and help when they analyze the observations to develop their evidence.

Topic Connections

Table 14.3 provides an overview of the scientific practices, crosscutting concepts, disciplinary core ideas, and supporting ideas at the heart of this lab investigation. In addition, it lists NOS and NOSI concepts for the explicit and reflective discussion. Finally, it lists literacy and mathematics skills (*CCSS ELA* and *CCSS Mathematics*) that are addressed during the investigation.

TABLE 14.3

Lab 14 alignment with standards

Scientific practices	• Asking questions and defining problems • Planning and carrying out investigations • Analyzing and interpreting data • Constructing explanations • Engaging in argument from evidence • Obtaining, evaluating, and communicating information
Crosscutting concepts	• Patterns • Structure and function
Core idea	• LS3: Heredity: Inheritance and variation in traits
Supporting ideas	• Species • Variation in traits • Comparing differences and similarities • Taxonomy
NOS and NOSI concepts	• Observations and inferences • Changes in scientific knowledge over time
Literacy connections (CCSS ELA)	• *Reading:* Key ideas and details, craft and structure, integration of knowledge and ideas • *Writing:* Text types and purposes, production and distribution of writing, research to build and present knowledge, range of writing • *Speaking and listening:* Comprehension and collaboration, presentation of knowledge and ideas
Mathematics connections (CCSS Mathematics)	• Make sense of problems and persevere in solving them • Construct viable arguments and critique the reasoning of others

LAB 14

Lab 14. Variation in Traits: How Do Beetle Traits Vary Within and Across Species?

Introduction

Organisms differ from one another in several ways. When those differences are so great that organisms are unable to mate and produce fertile offspring, they are said to be of different *species*. Differences between species, or so-called *interspecies differences*, can be great or small. Some differences can be easily seen, such as different shapes of arms or legs. Others are not easy to observe, such as differences in the kinds of genes each species has. Differences also exist within a species; in other words, not all members of the same species are the same—for example, not all gray whales are the same. Indeed, just look around you. All of the students in this class look very different even though they are all members of the same species. Some of these differences are unique to individual organisms or a group of organisms, especially in some physical traits, like special markings. Other differences, like ones in types of genes or behaviors, are more common among a particular species but make them distinct from other species. Some differences among species help certain forms survive better in certain environments than others.

These differences and similarities, both within and across species, are known as *variation*. To understand the variation present in living things, scientists developed a system of classification—that is, grouping organisms together based on their differences and similarities. This system begins with a few large groups of organisms that share a few similar traits on the first level. The next level of classification splits those large groups into smaller groups using differences and similarities among the organisms in them. This second set of groups is then split up even further based on the variation among the organisms. Current classification systems use eight major levels to classify all the living things. Some of these levels also have sublevels, which further improves classification. *Taxonomy* is the branch of science that deals with classifying organisms.

Take the beetle as an example of this classification system. The name Beetle is given to an order of insects that all have elytra. The elytra are a hardened, sheathlike set of front wings, which usually cover the entire abdomen when the insect is not in flight. Beetles vary greatly in size. The largest is the Eastern Hercules beetle (*Dynastes tityus*), which grows to 16 cm (up to 6.3 inches) in length (Figure L14.1). Other species may be less than 0.1 cm (less than 0.04 inch) long.

The Beetle order is split into a number of smaller groups, which involve more similarities in shape between the group members. These smaller groups follow a strict hierarchy. The major levels and sublevels smaller than order are called suborder, family, subfamily,

Eastern Hercules beetle

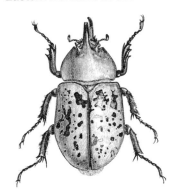

European violet ground beetle

and genus. A genus is the smallest group of importance in the naming of individual species, although in some classifications generic groups may be further split into subgenera. The scientific name of a species includes, first, its genus and, second, its specific name. For example, the European violet ground beetle (Figure L14.2) is called *Carabus violaceus*, meaning the species *violaceus* in the genus *Carabus*. (Genus and species names are always printed in italics. The genus name is capitalized, and the species name is lowercase [starts with a small letter].) The full classification of this insect is shown in Table L14.1.

TABLE L14.1 _____
Classification of the European violet ground beetle

Classification level	Name	Description
Phylum	Arthropoda	Arthropod
Class	Insecta	Insect
Subclass	Pterygota	Winged insect
Order	Coleoptera	Beetle
Suborder	Adephaga	Carnivorous beetle
Family	Carabidae	Ground beetle
Genus	*Carabus*	*
Species	*violaceus*	Violet ground beetle

The Beetle order embraces more species than any other group in the animal kingdom. At least 250,000 species are known. This represents more than one-quarter of all animal species and is far more than all mammal species combined. With all these different species, a lot of variation exists among beetles.

LAB 14

Your Task

Examine the amount and nature of the variation that exists within and across the three different species of beetle shown in the Lab Reference Sheet: The three species you can examine in the information sheets are the ground beetle (*Harpalus affinis*), the figeater beetle (*Cotinis mutabilis*), and the potato beetle (*Leptinotarsa decemlineata*).

The guiding question of this investigation is, **How do beetle traits vary within and across species?**

Materials

You may use any of the following materials during your investigation:

- Lab 14 Reference Sheet with beetle information, or preserved samples of beetles
- Insect information websites such as *www.bugguide.net* and *http://species.wikimedia.org*

Safety Precautions

Follow all normal lab safety rules.

Investigation Proposal Required?   ☐ Yes   ☐ No

Getting Started

To answer the guiding question, you will need to observe variations between different examples of beetles within each species and across the species provided. To accomplish this task, you must first determine what type of data you need to collect, how you will collect it, and how you will analyze it. To determine *what type of data you will need to collect*, think about the following questions:

- What type of data can you collect from the Lab Reference Sheet?
- What information about the beetles are available from internet sources?
- What type of measurements or observations will you need to record during your investigation?

To determine *how you will collect your data*, think about the following questions:

- What will serve as a control (or comparison) condition?
- How will you make sure that your data are of high quality (i.e., how will you reduce error)?
- How will you keep track of the data you collect and how will you organize the data?

To determine *how you will analyze your data*, think about the following questions:

- What type of calculations will you need to make?
- What type of graph could you create to help make sense of your data?

Connections to Crosscutting Concepts, the Nature of Science, the Nature of Scientific Inquiry

As you work through your investigation, be sure to think about

- the importance of looking for patterns in science,
- how organisms' structures are related to the functions they perform,
- the different roles observations and inferences play in science, and
- how scientific knowledge changes over time.

Initial Argument

Once your group has finished collecting and analyzing your data, you will need to develop an initial argument. Your argument must include a claim, evidence to support your claim, and a justification of the evidence. The claim is your group's answer to the guiding question. The evidence is an analysis and interpretation of your data. Finally, the justification of the evidence is why your group thinks the evidence matters. The justification of the evidence is important because scientists can use different kinds of evidence to support their claims. Your group will create your initial argument on a whiteboard. Your whiteboard should include all the information shown in Figure L14.3.

Argumentation Session

The argumentation session allows all of the groups to share their arguments. One member of each group will stay at the lab station to share that group's argument, while the other members of the group go to the other lab stations one at a time to listen to and critique the arguments developed by their classmates. This is similar to how scientists present their arguments to other scientists at conferences. If you are responsible for critiquing your classmates' arguments, your goal is to look for mistakes so these mistakes can be fixed and they can make their argument better. The argumentation session is also a good time to think about ways you can make your initial argument better. Scientists must share and critique arguments like this to develop new ideas.

To critique an argument, you might need more information than what is included on the whiteboard. You will therefore need to ask the presenter lots of questions. Here are some good questions to ask:

FIGURE L14.3

Argument presentation on a whiteboard

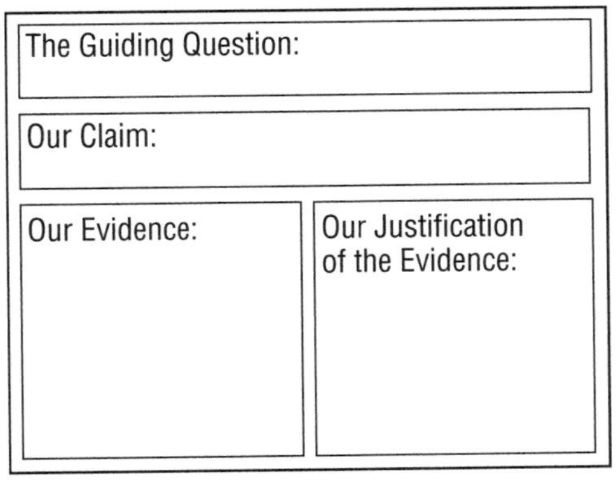

The Guiding Question:	
Our Claim:	
Our Evidence:	Our Justification of the Evidence:

- What did your group do to collect the data? Why do you think that way is the best way to do it?
- What did your group do to analyze the data? Why did your group decide to analyze it that way?
- What other ways of analyzing and interpreting the data did your group talk about?
- Why did your group decide to present your evidence in that way?
- What other claims did your group discuss before you decided on that one? Why did your group abandon those other ideas?
- How sure are you that your group's claim is accurate? What could you do to be more certain?

Once the argumentation session is complete, you will have a chance to meet with your group and revise your original argument. Your group might need to gather more data or design a way to test one or more alternative claims as part of this process. Remember, your goal at this stage of the investigation is to develop the most valid or acceptable answer to the research question!

Report

Once you have completed your research, you will need to prepare an investigation report that consists of three sections that provide answers to the following questions:

1. What question were you trying to answer and why?

2. What did you do during your investigation and why did you conduct your investigation in this way?

3. What is your argument?

Your report should answer these questions in two pages or less. The report must be typed, and any diagrams, figures, or tables should be embedded into the document. Be sure to write in a persuasive style; you are trying to convince others that your claim is acceptable or valid!

Lab 14 Reference Sheet
Three Types of Beetles

Ground beetle (*Harpalus affinis*)

a.

b.

c.

- Locations: Europe, Asia, North America, Middle East, and Australia
- Size: 8.5–12 mm
- Habitats: dry areas such as open farmland, parks, gardens, and sand dunes
- Reproduction
  - Spring is the main egg-laying season, but some egg laying occurs in summer.
  - Larvae and adults are present from winter into spring.

LAB 14

Figeater beetle (*Cotinis mutabilis*)

a.

b.

c.

- Locations: southwest United States, Mexico
- Size: 3 cm (adult)
- Habitats: areas where they can feed on pollen and nectar, as well as damaged fruits
- Reproduction:
    - Spring is the main time for transition from larvae to adults.
    - Adults emerge July–September.
    - Eggs and larvae stage occur over winter.

Potato beetle (*Leptinotarsa decemlineata*)

a.

b.

c.

- Locations: southwest United States, Mexico, Europe, and southern Russia
- Size: 10 mm (adult)
- Habitats: found mainly near farmland (crop pest for potato agriculture)
- Reproduction
  - Very quick, typically one month to go from egg to adult
  - Eggs laid on the underside of leaves
  - Temperature and light dependent
  - Three generations can grow during a single crop season

LAB 14

Lab 14. Variation in Traits: How Do Beetle Traits Vary Within and Across Species?

1. Describe how variation exists among different species that are similar.

2. Physical and genetic differences exist among the many types of living organisms that exist on the planet. Explain how scientists use the variation they observe among different species to classify them in different size groups. Does more variation exist among organisms that are in the same class or the same family? Explain your answer.

3. Observations are the same thing in science as inferences.

   a. I agree with this statement.
   b. I disagree with this statement.

   Explain your answer, using an example from your investigation about variation in traits.

4. Scientific knowledge can change over time but is still reliable.

   a. I agree with this statement.
   b. I disagree with this statement.

   Explain your answer, using an example from your investigation about variation in traits.

5. Scientists often look for patterns when they study nature. Explain why understanding patterns in nature are important, using an example from your investigation about variation in traits.

6. The relationship between the structure and function of organisms' features is an important area of study in science. Discuss why it is important for scientists to understand this relationship, using an example from your investigation about variations in traits.

LAB 15

Teacher Notes

Lab 15. Mutations in Genes: How Do Different Types of Mutations in Genes Affect the Function of an Organism?

Purpose

The purpose of this lab is to *introduce* students to the transfer of genetic information through inheritance and how DNA-level mutations can influence the functions of organisms. Specifically, this investigation gives students an opportunity to use a computer simulation to explore how different nucleotide (sugar/phosphate + base = smallest unit of DNA) mutations affect the shape of proteins made from a sequence of DNA. Teachers will need to help students understand how the structure and function of DNA influences the structure and function of proteins. This lab gives students an opportunity to explore the importance of scales and proportions when studying scientific phenomena. Students will also learn about the connections between structure and functions when studying living things. Students will have the opportunity to reflect on the different roles for theories and laws in science and how imagination and creativity are necessary characteristics in developing scientific knowledge.

The Content

In the mid-1800s, Gregor Mendel investigated how certain traits in pea plants were transferred from one generation of plants to the next. He figured out that the pea plants had "factors" that carried the information for those traits. During reproduction, those factors transferred information for traits from the parent plants to their offspring. This transfer of trait information from parents to offspring is known as *inheritance*. He continued his investigations to figure out several patterns in the way these factors moved from one generation to the next. Through more work by many other scientists over the years, more is known about Mendel's factors involved in inheritance. These factors, which are found in all living things, are called *genes*. Indeed, reproduction in living things mainly deals with the transfer of genes from parents to offspring. All of the genes passed on during reproduction contain information needed to create a new organism. There can also be different versions of the same gene, like flower color (purple or white) or plant height (tall or short) in Mendel's pea plants. These versions are called *alleles*.

Using the patterns he observed, Mendel developed several scientific laws of inheritance, including the law of segregation, the law of independent assortment, and the law of dominance. These laws tell us about the patterns in which information transferred during inheritance moves from parent to offspring during reproduction (segregation and independent assortment) and how that information is expressed (dominance). Remember that scientific laws describe relationships between two components of a system. The law of segregation describes how the two copies of a gene present in an organism separate during cell division that forms gametes. The law of independent assortment states that separate genes for different traits are passed on independently of each other from parent to offspring, not influencing the inheritance of each other. The law of dominance states that, in most cases, when two different alleles for the same gene are brought together in an organism's body cells, the trait produced by one version (recessive) will be masked by the other version (dominant), which produces the expressed trait.

Following Mendel's work, several other important scientists (Theodor Boveri, Elinor Carothers, Thomas Hunt Morgan, Walter Sutton) developed major findings from their research that contributed to the *chromosomal theory of inheritance*. Remember that theories serve as broad explanations for complex phenomena that bring together multiple lines of evidence in support. The chromosomal theory of inheritance describes how genes are the factors that determine traits, which are carried on larger structures in an organism's cell nuclei, called chromosomes. These chromosomes are transferred through gametes during reproduction and are fixed types for different species.

Genes are made of a molecule known as *DNA*, which stands for deoxyribonucleic acid Figure 15.1 shows a section of a DNA molecule. DNA is made of two strands of molecules with a sugar/phosphate side and a base side. The sugar/phosphate sides of single molecules bond together to form the chains. The base sides interact with the base sides of another strand to connect the two strands together. However, there are only four types of bases in DNA: adenine (A), guanine (G), cytosine (C), and thymine (T). These four bases will bond in only two ways: an A on one strand will only bond to a T on the other strand, and a C on one strand will only bond to a G on the other strand. As these bases pair up, the two strands stay connected, forming larger molecules of DNA. DNA molecules naturally twist to form a *double helix*, which is shaped like a twisted ladder. This understanding of the structure of DNA and its role in storing and transferring genetic information was made possible through the work of many scientists, including Alfred Hershey, Martha Chase, Erwin Chargaff, Francis Crick, Rosalind Franklin, James Watson, and Maurice Wilkins.

The chemical structure of the DNA molecule allows it to store information. The order of the bases in a strand contains a code for the structure of other necessary molecules in organisms. When cells reproduce, they must copy the DNA inside of them. The order of the bases in the DNA is used to make copies of that DNA. This copying allows parents to pass on copies of their genes to their offspring. A section of a strand of DNA, like TACCGATGATTCCGG, also has a code that can tell an organism's cells how to build other molecules. Genes are made up of long units of DNA.

FIGURE 15.1 _____
Section of a DNA molecule

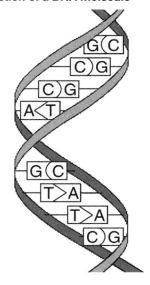

LAB 15

The DNA in genes is read by special molecules in cells called *enzymes* that use the code in DNA to build an *RNA* molecule. RNA, which stands for ribonucleic acid, is a single-stranded molecule similar to DNA that also contains a sequence of bases. The process of building RNA from DNA is called *transcription*. The RNA molecule made from a specific gene is then used to make a *protein* molecule (see the next paragraph for a discussion of proteins). The process of making a protein molecule from a RNA molecule is called *translation*. Figure 15.2 shows the connections between DNA, RNA, and proteins. This diagram is also a simplified version of the overarching theory for molecular biology, commonly known as the *central dogma*. However, this theory is still in development as more discoveries about the behavior of DNA, RNA, and proteins are made. Recently, scientists have determined that RNA is used to make DNA, especially certain types of sequences. Also, scientists continue to discover different types of RNA that appear to serve very important functions not related to the making of proteins. The central dogma will continue to develop as technology increases scientists' investigative abilities and they develop more complex understandings of the interactions between these molecules.

FIGURE 15.2

The central dogma

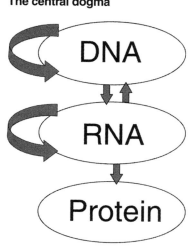

Proteins are molecules that perform all kinds of functions in cells. These functions include moving other molecules around an organism's body and forming the muscles that help bodies move. A protein's function is determined by the way it is shaped. That shape is determined by the order of *amino acids* that make up a protein's structure. Amino acids are small molecules that make up larger protein molecules, much like the sugar/phosphate + base nucleotides that make up DNA and RNA. The order of the amino acids determines how those chains will fold in on each other to form the larger three-dimensional (3-D) shape of the protein. The 3-D shape of a protein is critical in allowing proteins to interact or react with other chemical molecules and structures that they are supposed to in order to function properly and support life. Thus, in all of these large molecules, the order structure of the smaller molecules that form them determines the functions the larger molecules perform.

Changes to the order of those small molecules are called *mutations*. Mutations can occur naturally or can be caused by interactions with other chemicals. Mutations are often thought of in a negative and harmful way. However, genetic mutation is the fundamental source for biological evolution. Without mutation, the variation necessary for creating reproductive advantages in species would not be possible. There are several types of mutations that can occur to the order of bases in DNA, as illustrated in Figure 15.3 using the same original DNA sequence:

- *Substitution mutations* happen when pairs of bonded bases in a double strand of DNA get replaced with different pairs. An example of a substitution mutation is having an A-T base pair replaced by a G-C pair or a T-A pair.

FIGURE 15.3

Types of DNA-level mutations

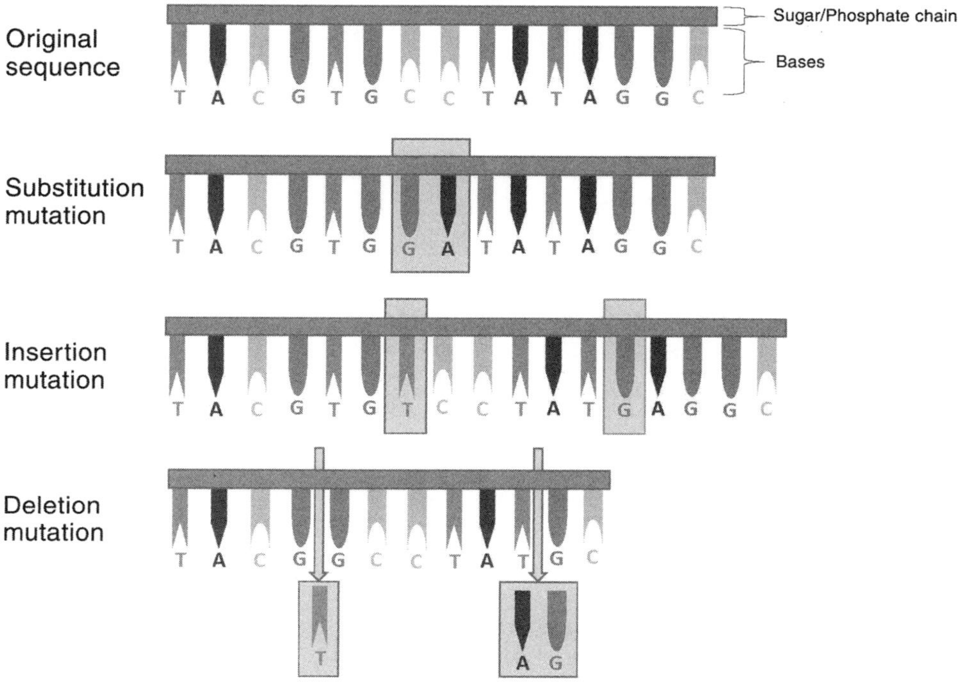

- *Insertion mutations* happen when extra base pairs are included in an existing DNA sequence.
- *Deletion mutations* happen when base pairs in an existing DNA molecule get removed from it.

For all of these mutations, the location in a coding sequence of DNA is important. Some mutations will not change the amino acid coded for by a specific sequence, but others will. If the amino acid changes, in some cases the chemical nature is so similar that it does not affect the 3-D shape of the protein. However, in other cases it can completely rearrange the 3-D shape. These mutations are the major focus for this lab activity, however, they are not the only type of genetic mutation possible. There are also several types of chromosomal-level mutations that can have serious impacts on how organisms develop and function.

Timeline

The instructional time needed to implement this lab investigation is 180–250 minutes. Appendix 2 (p. 355) provides options for implementing this lab investigation over several class periods. Option A (250 minutes) should be used if students are unfamiliar with scientific writing, because this option provides extra instructional time for scaffolding the writing process. You can scaffold the writing process by modeling, providing examples, and providing hints as students write each section of the report. Option B (180 minutes) should be used if students are familiar with scientific writing and have the skills needed to write an investigation report on their own. In option B, students complete stage 6 (writing the investigation report) and stage 8 (revising the investigation report) as homework.

Materials and Preparation

The materials needed to implement this investigation are listed in Table 15.1. The *Mutations* simulation, available at *http://concord.org/stem-resources/mutations*, is free to use and can be run online using an internet browser. You can use a computer lab or a classroom set of laptops. You should access the website and learn how the simulation works before beginning the lab investigation. In addition, it is important to check if students can access and use the simulation from a school computer, because some schools have set up firewalls and other restrictions on web browsing.

TABLE 15.1
Materials list

Item	Quantity
Computer with internet access	1 per group
Investigation Proposal C (optional)	1 per group
Whiteboard, 2' × 3'*	1 per group
Lab Handout	1 per student
Peer-review guide	1 per student
Checkout Questions	1 per student

* As an alternative, students can use computer and presentation software such as Microsoft PowerPoint or Apple Keynote to create their arguments.

Safety Precautions

Follow all normal lab safety rules.

Topics for the Explicit and Reflective Discussion

Concepts That Can Be Used to Justify the Evidence

To provide an adequate justification of their evidence, students must explain why they included the evidence in their arguments and make the assumptions underlying their analysis and interpretation of the data explicit. In this investigation, students can use the following concepts to help justify their evidence:

- Mutations change the structure of molecules, including DNA, RNA, and proteins.
- The sequence of bases in DNA determines the order of amino acids in a protein.
- The structure of a molecule determines the functions it can perform.
- Some mutations will not change the amino acid sequence of the protein, having less impact than others.

We recommend that you review these concepts during the explicit and reflective discussion to help students make this connection.

How to Design Better Investigations

It is important for students to reflect on the strengths and weaknesses of the investigation they designed during the explicit and reflective discussion. Students should therefore be encouraged to discuss ways to eliminate potential flaws, measurement errors, or sources of bias in their investigations. To help students be more reflective about the design of their investigation, you can ask the following questions:

- What were some of the strengths of your investigation? What made it scientific?
- What were some of the weaknesses of your investigation? What made it less scientific?
- If you were to do this investigation again, what would you do to address the weaknesses in your investigation? What could you do to make it more scientific?

Crosscutting Concepts

This investigation is aligned with two crosscutting concepts found in *A Framework for K–12 Science Education*, and you should review these concepts during the explicit and reflective discussion.

- *Scale, proportion, and quantity:* It is critical for scientists to be able to recognize what is relevant at different sizes, time frames, and scales. Scientists must also be able to recognize proportional relationships between categories, groups, or quantities.
- *Structure and function:* In nature, the way a living thing is structured or shaped determines how it functions and places limits on what it can and cannot do.

LAB 15

The Nature of Science and the Nature of Scientific Inquiry

This investigation is aligned with two important concepts related to the *nature of science* (NOS) and the *nature of scientific inquiry* (NOSI), and you should review these concepts during the explicit and reflective discussion.

- *The difference between laws and theories in science:* A scientific law describes the behavior of a natural phenomenon or a generalized relationship under certain conditions; a scientific theory is a well-substantiated explanation of some aspect of the natural world. Theories do not become laws even with additional evidence; they explain laws. However, not all scientific laws have an accompanying explanatory theory. It is also important for students to understand that scientists do not discover laws or theories; the scientific community develops them over time.

- *The importance of imagination and creativity in science:* Students should learn that developing explanations for or models of natural phenomena and then figuring out how they can be put to the test of reality is as creative as writing poetry, composing music, or designing skyscrapers. Scientists must also use their imagination and creativity to figure out new ways to test ideas and collect or analyze data.

Hints for Implementing the Lab

- Have students run each type of mutation several times. Also, make sure they run the simulation without any mutations included so they can see the changes the mutations have on the folded protein.

- Emphasize the folded shape of the protein at the end of each simulation as a major observation/data point the students should use.

Topic Connections

Table 15.2 provides an overview of the scientific practices, crosscutting concepts, disciplinary core ideas, and supporting ideas at the heart of this lab investigation. In addition, it lists NOS and NOSI concepts for the explicit and reflective discussion. Finally, it lists literacy and mathematics skills (*CCSS ELA* and *CCSS Mathematics*) that are addressed during the investigation.

TABLE 15.2

Lab 15 alignment with standards

Scientific practices	• Asking questions and defining problems • Developing and using models • Planning and carrying out investigations • Analyzing and interpreting data • Constructing explanations • Engaging in argument from evidence • Obtaining, evaluating, and communicating information
Crosscutting concepts	• Scale, proportion, and quantity • Structure and function
Core idea	• LS3: Heredity: Inheritance and Variation of Traits
Supporting ideas	• Mutation • DNA, RNA, and proteins • Molecular structure and function
NOS and NOSI concepts	• Scientific laws and theories • Imagination and creativity in science
Literacy connections (CCSS ELA)	• *Reading:* Key ideas and details, craft and structure, integration of knowledge and ideas • *Writing:* Text types and purposes, production and distribution of writing, research to build and present knowledge, range of writing • *Speaking and listening:* Comprehension and collaboration, presentation of knowledge and ideas
Mathematics connections (CCSS Mathematics)	• Make sense of problems and persevere in solving them • Reason abstractly and quantitatively • Construct viable arguments and critique the reasoning of others • Use appropriate tools strategically

Reference

The Concord Consortium. 2015. Mutations. Concord, MA: The Concord Consortium. *http://concord.org/stem-resources/mutations.*

LAB 15

Lab 15. Mutations in Genes: How Do Different Types of Mutations in Genes Affect the Function of an Organism?

FIGURE L15.1

Section of a DNA molecule

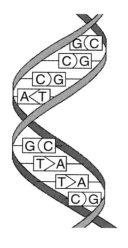

FIGURE L15.2

Connections between DNA, RNA, and proteins

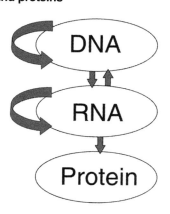

Introduction

During reproduction, information for traits is passed from the parent organisms to the offspring. This transfer of trait information from parents to offspring is known as *inheritance*. The factors with the trait information are found in all living things and are called *genes*. All of the genes passed on during reproduction contain information needed to create a new organism. Genes are made of a molecule known as *DNA*, which stands for deoxyribonucleic acid. Figure L15.1 shows a section of a DNA molecule. DNA is made of two strands of molecules with a sugar/phosphate side and a base side. The base sides interact with the base sides of another strand to connect the two strands together. There are only four types of bases in DNA, called *A*, *G*, *C*, and *T*. These four bases will bond in only two ways: an A on one strand will only bond to a T on the other strand, and a C on one strand will only bond to a G on the other strand. As these bases pair up, the two strands stay connected, forming larger molecules of DNA.

The chemical structure of the DNA molecule allows it to store information. Genes are made up of long units of DNA. The order of the bases in a strand contains a code for the structure of other molecules in organisms. A section of a strand of DNA, like TACCGATGATTCCGG, has a code that tells an organism's cells how to build other molecules. The DNA in genes is read by special molecules, called enzymes, which use the code in DNA to build an *RNA* molecule. RNA, which stands for ribonucleic acid, is a single-stranded molecule similar to DNA that also contains a sequence of bases. The RNA molecule made from a specific gene is then used to make a *protein* molecule (see the next paragraph). Figure L15.2 shows the connections between DNA, RNA, and proteins.

Proteins are molecules that perform all kinds of functions in cells. A protein's function is determined by the way it is shaped, and that shape is determined by the order of *amino acids* that make up a protein's structure. Amino acids are small molecules that make up larger protein molecules, much like the sugar/phosphate + base molecules make up DNA and RNA. Thus, in all of these large molecules, the

Mutations in Genes

How Do Different Types of Mutations in Genes Affect the Function of an Organism?

order structure of the smaller molecules that form them determine the functions the larger molecules perform.

Changes to the order of those small molecules are called *mutations*. There are several types of mutations that can occur to the order of bases in DNA. *Substitution mutations* happen when pairs of bonded bases in a double strand of DNA get replaced with different pairs. An example of a substitution mutation is having an A-T base pair replaced by a G-C pair or a T-A pair. *Insertion mutations* happen when extra base pairs are included in an existing DNA sequence. *Deletion mutations* happen when base pairs in an existing DNA molecule get removed from it. Figure L15.3 shows each type of mutation using the same original DNA sequence. For all of these mutations, the location in the order of bases in DNA is important. Some mutations will not change the amino acid coded for by a specific sequence, but others will.

FIGURE L15.3

Types of DNA-level mutations

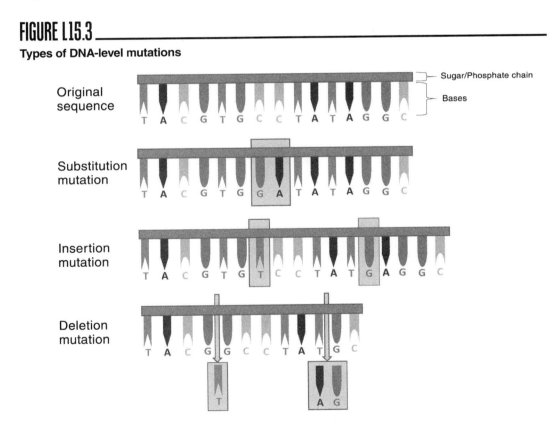

LAB 15

Your Task

Examine the effects of different types of mutations in a DNA sequence on the resulting RNA and protein molecules. Since these molecules are not easy to work with in a classroom, you will be using a computer simulation to investigate the effect of mutations.

The guiding question of this investigation is, **How do different types of mutations in genes affect the function of an organism?**

Materials

You will use an online simulation called *Mutations* to conduct your investigation. You can find the simulation by going to the following website: *http://concord.org/stem-resources/mutations*.

Safety Precautions

Follow all normal lab safety rules.

Investigation Proposal Required?   ☐ Yes   ☐ No

Getting Started

To answer the guiding question, you will use a simulation to observe the impact of different mutations on the resulting protein. To accomplish this task, you must first determine what type of data you need to collect, how you will collect it, and how you will analyze it. To determine *what type of data you will need to collect*, think about the following questions:

- What type of data can you collect from the simulation?
- What types of mutations can you make using the simulation?
- What type of measurements or observations will you need to record during your investigation?

To determine *how you will collect your data*, think about the following questions:

- What will serve as a control (or comparison) condition?
- How will you make sure that your data are of high quality (i.e., how will you reduce error)?
- How will you keep track of the data you collect and how will you organize the data?

To determine *how you will analyze your data*, think about the following questions:

- What type of calculations will you need to make?
- What type of graph could you create to help make sense of your data?

Connections to Crosscutting Concepts, the Nature of Science, and the Nature of Scientific Inquiry

As you work through your investigation, be sure to think about

- how different actions in living things happen on different scales of size and time,
- how organisms' structures are related to the functions they perform,
- the different roles theories and laws play in science, and
- how imagination and creativity are necessary for developing scientific knowledge.

Initial Argument

Once your group has finished collecting and analyzing your data, you will need to develop an initial argument. Your argument must include a claim, evidence to support your claim, and a justification of the evidence. The claim is your group's answer to the guiding question. The evidence is an analysis and interpretation of your data. Finally, the justification of the evidence is why your group thinks the evidence matters. The justification of the evidence is important because scientists can use different kinds of evidence to support their claims. Your group will create your initial argument on a whiteboard. Your whiteboard should include all the information shown in Figure L15.4.

FIGURE L15.4

Argument presentation on a whiteboard

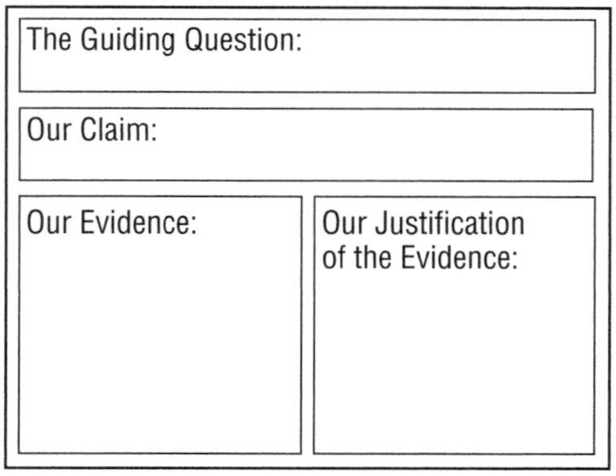

The Guiding Question:	
Our Claim:	
Our Evidence:	Our Justification of the Evidence:

Argumentation Session

The argumentation session allows all of the groups to share their arguments. One member of each group will stay at the lab station to share that group's argument, while the other members of the group go to the other lab stations one at a time to listen to and critique the arguments developed by their classmates. This is similar to how scientists present their arguments to other scientists at conferences. If you are responsible for critiquing your classmates' arguments, your goal is to look for mistakes so these mistakes can be fixed and they can make their argument better. The argumentation session is also a good time to think about ways you can make your initial argument better. Scientists must share and critique arguments like this to develop new ideas.

To critique an argument, you might need more information than what is included on the whiteboard. You will therefore need to ask the presenter lots of questions. Here are some good questions to ask:

- What did your group do to collect the data? Why do you think that way is the best way to do it?

- What did your group do to analyze the data? Why did your group decide to analyze it that way?

- What other ways of analyzing and interpreting the data did your group talk about?

- Why did your group decide to present your evidence in that way?

- What other claims did your group discuss before you decided on that one? Why did your group abandon those other ideas?

- How sure are you that your group's claim is accurate? What could you do to be more certain?

Once the argumentation session is complete, you will have a chance to meet with your group and revise your original argument. Your group might need to gather more data or design a way to test one or more alternative claims as part of this process. Remember, your goal at this stage of the investigation is to develop the most valid or acceptable answer to the research question!

Report

Once you have completed your research, you will need to prepare an investigation report that consists of three sections that provide answers to the following questions:

1. What question were you trying to answer and why?

2. What did you do during your investigation and why did you conduct your investigation in this way?

3. What is your argument?

Your report should answer these questions in two pages or less. The report must be typed, and any diagrams, figures, or tables should be embedded into the document. Be sure to write in a persuasive style; you are trying to convince others that your claim is acceptable or valid!

Checkout Questions

Lab 15. Mutations in Genes: How Do Different Types of Mutations in Genes Affect the Function of an Organism?

1. Describe the different kinds of mutations that can happen in a sequence of DNA.

2. A muscle cell in an organism's body has stopped making one of the proteins needed to make a muscle fiber. That cell was exposed to a large amount of radiation previously. Explain how mutations could have led to this problem in this muscle cell.

3. Theories and laws serve the same purpose in science.

   a. I agree with this statement.
   b. I disagree with this statement.

   Explain your answer, using an example from your investigation about mutations in genes.

4. Creativity is an important characteristic for a scientist to have.

    a. I agree with this statement.

    b. I disagree with this statement.

    Explain your answer, using an example from your investigation about mutations in genes.

5. Scientists often observe events that happen over different time scales; some events are very quick and others take many years. Explain why understanding different time scales is important, using an example from your investigation about mutations in genes.

6. The relationship between the structure and function of organisms' features is an important area of study in science. Discuss why it is important for scientists to understand this relationship, using an example from your investigation about mutations in genes.

Application Lab

LAB 16

Lab 16. Mechanisms of Inheritance: How Do Fruit Flies Inherit the Sepia Eye Color Trait?

Purpose

The purpose of this lab is for students to *apply* what they know about the inheritance of traits to solve a problem. Specifically, this investigation gives students an opportunity to use several conceptual models—the dominant-recessive, incomplete dominance, and codominant models of inheritance—as a way to make sense of a natural phenomenon. This lab also gives students an opportunity to learn how scientists use patterns and models to understand complex phenomena. Students will have the opportunity to reflect on the differences between theories and laws and between data and evidence.

The Content

The principles of Mendelian genetics encompass several different models of inheritance. These models include dominant-recessive, incomplete dominance, and codominance. All of these models are based on two fundamental ideas. The first and most important idea is the gene. The gene is the fundamental unit of inheritance, and alternative versions of a gene (*alleles*) account for variations in inheritable traits. The gene for a particular inherited trait, such as eye color in fruit flies, resides at a specific locus (position) on a specific chromosome. Second, an organism inherits two alleles for each trait, one from each parent. This occurs because individuals inherit one chromosome for each homologous pair from each parent. What makes these models of inheritance different from each other is the nature of the interaction that takes place between alleles, the number of alleles that are associated with a gene, and whether or not a gene is located on a sex chromosome.

The *dominant-recessive model of inheritance* is the model developed by Gregor Mendel. In this model, an individual inherits two alleles. When the two alleles differ, one is fully expressed and determines the nature of the trait (this version of the gene is called the dominant allele) while the other one has no noticeable effect (this version of the gene is called the recessive allele).

In contrast to the dominant-recessive model, the *incomplete dominance model of inheritance* suggests that alleles interact with each other to produce a trait that is a hybrid phenotype of the two parental varieties. A well-known example of incomplete dominance in humans is hair texture. When an individual inherits the allele for curly hair from one parent and the allele for straight hair from the other parent, that individual will have a hair texture that is a blend of the two, or wavy hair.

In the *codominance model of inheritance,* both alleles affect the phenotype of the individual in separate and distinguishable ways; the coat color of Shorthorn cattle is an example of this model. Shorthorn cattle that are homozygous for the R allele have red hair, and individuals that are homozygous for the W allele have white hair. Individuals that inherit a copy of the R allele and the W allele have intermingled red and white hair (not pink hair).

Timeline

The instructional time needed to implement this lab investigation is 180–250 minutes. Appendix 2 (p. 355) provides options for implementing this lab investigation over several class periods. Option A (250 minutes) should be used if students are unfamiliar with scientific writing, because either of these options provides extra instructional time for scaffolding the writing process. You can scaffold the writing process by modeling, providing examples, and providing hints as students write each section of the report. Option B (180 minutes) should be used if students are familiar with scientific writing and have the skills needed to write an investigation report on their own. In option B, students complete stage 6 (writing the investigation report) and stage 8 (revising the investigation report) as homework.

Materials and Preparation

The materials needed to implement this investigation are listed in Table 16.1 (p. 266). The *Drosophila* simulation, created by the Virtual Courseware for Inquiry-Based Science Education (VCISE) project at California State University, Los Angeles (see *www.sciencecourseware.org/vcise*), is free to use and can be run online using an internet browser. You should access the website and learn how the simulation works before beginning the lab investigation. In addition, it is important to check if students can access and use the simulation from a school computer, because some schools have set up firewalls and other restrictions on web browsing.

There are two options for using this simulation: (1) you can have students use the simulation as a guest or (2) you can create an account and be given access codes for each class. The access codes allow students to save their work. If you choose to create a login for yourself and your classes, you will need to follow this procedure:

1. Go to the VCISE website (*www.sciencecourseware.org/vcise*) and create a free teacher account.

2. Log in to your new teacher account and register your classes (click "Add a New Class").

3. Provide the class code you are given to your students so they can use it when they register to use the simulation.

LAB 16

TABLE 16.1
Materials list

Item	Quantity
Computer with internet access	1 per group
Investigation Proposal A (optional)	1 per group
Whiteboard, 2' × 3'*	1 per group
Lab Handout	1 per student
Peer-review guide	1 per student
Checkout Questions	1 per student

* Students can also use computer and presentation software such as Microsoft PowerPoint or Apple Keynote to create their arguments.

Safety Precautions

Follow all normal lab safety rules.

Topics for the Explicit and Reflective Discussion

Concepts That Can Be Used to Justify the Evidence

To provide an adequate justification of their evidence, students must explain why they included the evidence in their arguments and make the assumptions underlying their analysis and interpretation of the data explicit. In this investigation, students must use the following concepts to help justify their evidence:

- Mendel's laws
- Dominant-recessive model of inheritance
- Incomplete dominance model of inheritance
- Codominance model of inheritance

We recommend that you review these concepts during the explicit and reflective discussion to help students make this connection.

How to Design Better Investigations

It is important for students to reflect on the strengths and weaknesses of the investigation they designed during the explicit and reflective discussion. Students should therefore be encouraged to discuss ways to eliminate potential flaws, measurement errors, or sources of bias in their investigations. To help students be more reflective about the design of their investigation, you can ask the following questions:

- What were some of the strengths of your investigation? What made it scientific?

- What were some of the weaknesses of your investigation? What made it less scientific?

- If you were to do this investigation again, what would you do to address the weaknesses in your investigation? What could you do to make it more scientific?

Crosscutting Concepts

This investigation is aligned with two crosscutting concepts found in *A Framework for K–12 Science Education,* and you should review these concepts during the explicit and reflective discussion.

- *Patterns:* Observed patterns in nature can help us understand natural phenomena and their underlying causes. In this investigation, for example, the students need to use the observed patterns in the ratio of offspring with specific traits after conducting a test cross to test their ideas about how a specific trait is inherited.

- *Systems and system models:* Scientists often need to use explanatory models to understand complex phenomena. In this investigation, for example, students need to use the various models of inheritance to interpret the results of the test crosses that they performed using the online simulation.

The Nature of Science and the Nature of Scientific Inquiry

This investigation is aligned with two important concepts related to the *nature of science* (NOS) and the *nature of scientific inquiry* (NOSI), and you should review these concepts during the explicit and reflective discussion.

- *The difference between laws and theories in science:* A scientific law describes the behavior of a natural phenomenon or a generalized relationship under certain conditions; a scientific theory is a well-substantiated explanation of some aspect of the natural world. Theories do not become laws even with additional evidence; they explain laws. However, not all scientific laws have an accompanying explanatory theory. It is also important for students to understand that scientists do not discover laws or theories; the scientific community develops them over time.

- *The difference between data and evidence in science:* Data are measurements, observations, and findings from other studies that are collected as part of an investigation. Evidence, in contrast, is analyzed data and an interpretation of the analysis.

LAB 16

Hints for Implementing the Lab

- Students should have a good understanding of the different models of inheritance before beginning this investigation. They should also know how to perform a test cross and how to set up and interpret a Punnett square.

- To learn more about the simulation, take the "tour of the simulation" when you log in as a guest or consult the simulation manual, which can be found on the tabs on the left-hand side of the window when you log in to a teacher account on the VCISE homepage (*www.sciencecourseware.org/vcise*).

- Remind students to keep detailed records of their data, and encourage the use of tables!

- If you have limited access to computers (either number or time allotted), have students familiarize themselves with the simulation by taking the simulation tour outside of class time. They can access the simulation tour at home or at a library if needed.

Topic Connections

Table 16.2 provides an overview of the scientific practices, crosscutting concepts, disciplinary core ideas, and support ideas at the heart of this lab investigation. In addition, it lists NOS and NOSI concepts for the explicit and reflective discussion. Finally, it lists literacy and mathematics skills (*CCSS ELA* and *CCSS Mathematics*) that are addressed during the investigation.

TABLE 16.2

Lab 16 alignment with standards

Scientific practices	• Asking questions and defining problems • Developing and using models • Planning and carrying out investigations • Analyzing and interpreting data • Using mathematics and computational thinking • Constructing explanations • Engaging in argument from evidence • Obtaining, evaluating, and communicating information
Crosscutting concepts	• Patterns • Systems and system models
Core ideas	• LS3: Heredity: Inheritance and variation of traits
Supporting ideas	• Dominant-recessive model of inheritance • Incomplete dominance model of inheritance • Codominance model of inheritance • Phenotypes • Test cross • Punnett squares
NOS and NOSI concepts	• Scientific laws and theories • Difference between data and evidence
Literacy connections (*CCSS ELA*)	• *Reading:* Key ideas and details, craft and structure, integration of knowledge and ideas • *Writing:* Text types and purposes, production and distribution of writing, research to build and present knowledge, range of writing • *Speaking and listening:* Comprehension and collaboration, presentation of knowledge and ideas
Mathematics connections (*CCSS Mathematics*)	• Make sense of problems and persevere in solving them • Reason abstractly and quantitatively • Construct viable arguments and critique the reasoning of others • Use appropriate tools strategically • Look for and make use of structure • Look for and express regularity in repeated reasoning

LAB 16

Lab Handout

Lab 16. Mechanisms of Inheritance: How Do Fruit Flies Inherit the Sepia Eye Color Trait?

Introduction

In the 1800s, farmers often cross-pollinated specific types of plants or mated livestock with specific traits in an effort to produce offspring with more desirable traits. Their selective breeding process, however, did not always lead to the desired outcome. These farmers were often unsuccessful in their attempts to produce offspring with specific traits because they did not understand the mechanisms that control how traits are passed down from parent to offspring. The inheritance of traits also baffled the scientists of that time, until Gregor Mendel was able to explain the rules that govern heredity in 1865.

Mendel was able to explain how and why specific traits are passed down from generation to generation by breeding pea plants. He first cross-pollinated individual pea plants with specific traits and documented the traits of their offspring. He then cross-pollinated individual offspring and documented the traits that shown up in the next generation. By doing this, he was able to identify patterns in the ways traits are passed down from one generation to the next. He then developed several rules that he could use to explain the patterns he uncovered in the ways traits are inherited; these rules are now called *Mendel's laws* and can be summarized as follows:

- Inheritable units called *genes* determine traits.
- A gene comes in different versions called *alleles*.
- An organism carries two alleles for each trait.
- The two alleles segregate during *gamete* (a reproductive cell such as a sperm or an egg) production, so each gamete only carries one allele for a specific trait.
- When gametes unite during fertilization, each contributes its allele, so offspring inherit one allele from each parent.

There are several different models of inheritance that are based on Mendel's laws. These models include dominant-recessive, incomplete dominance, and codominance. What makes these models of inheritance different from each other is how each one describes the interaction or behavior of the alleles once they have been passed down from parent to offspring. The *dominant-recessive model of inheritance* suggests that when an individual inherits two alleles and the two alleles differ, then one is fully expressed and determines the trait (this version of the gene is called the dominant allele) while the other one has no noticeable effect (this version of the gene is called the recessive allele). The *incomplete dominance model of inheritance* suggests that the interaction that occurs between two different alleles results in a hybrid with an appearance somewhere between the phenotypes of the two parental varieties. The *codominance model of inheritance* is similar to the incomplete dominance model, but in the codominance model both alleles affect the phenotype of the individual in separate and distinguishable ways.

It is often difficult, however, to determine which of these three models of inheritance best explains how a specific trait is inherited. To illustrate this point, you will be studying the inheritance of a trait in fruit flies (*Drosophila melanogaster*). Fruit flies are very common. Most fruit flies have six legs, two wings, and two antennae (see Figure L16.1). Most fruit flies also have an orange-yellow body and red eyes. Scientist call flies with these traits the "wild type." Every once in a while, however, you might see a fruit fly with sepia (brown) or white eyes. In this investigation, you will need to determine how the sepia eye color is inherited.

FIGURE L16.1

Male and female fruit flies

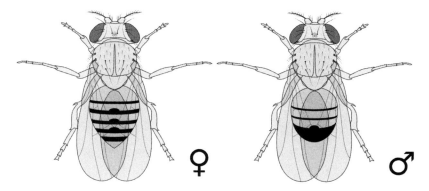

Your Task

Use a computer simulation that allows you to "order" fruit flies with specific traits from a supply company and then "breed" them to see how the sepia eye color trait is passed down from parent to offspring. From there, you will need to decide how the allele for sepia eye color interacts with other alleles for eye color (such as the allele for red eyes). You will then decide which model of inheritance (dominant-recessive, incomplete dominance, or codominance) best explains how the sepia eye color trait is inherited in fruit flies.

The guiding question of this investigation is, **How do fruit flies inherit the sepia eye color trait?**

LAB 16

Materials

You will use an online simulation called *Drosophila* to conduct your investigation. You can find the simulation by going to the following website: *www.sciencecourseware.org/ vcise/drosophila*.

Safety Precautions

Follow all normal lab safety rules.

Investigation Proposal Required?   ☐ Yes   ☐ No

Getting Started

Your teacher will show you how to use the *Drosophila* online simulation before you begin designing your investigation. The first step in the investigation is to learn how the sepia eye color trait is passed from one generation to the next. To accomplish this task, you must determine what type of data you need to collect using the online simulation, how you will collect these data, and how you will analyze the data. To determine *what type of data you will need to collect*, think about the following questions:

- What types of flies will you need to work with during your investigation (e.g., males or females, flies with red eyes or flies with sepia eyes)?
- What type of measurements or observations will you need to record during your investigation?
- How will you identify a pattern in the way the sepia eye color trait is inherited?

To determine *how you will collect your data*, think about the following questions:

- How many times will you need to breed the flies?
- How many generations of flies will you need to follow?
- How often will you collect data and when will you do it?
- How will you keep track of the data you collect and how will you organize the data?

To determine *how you will analyze your data*, think about the following questions:

- How will you determine if the results of your cross tests match your predictions?
- What type of graph could you create to help make sense of your data?

The last step in this investigation is to test the model of inheritance that you think best explains the inheritance of the sepia eye color trait. To accomplish this goal, you can use the simulation to determine if you can use your model to make accurate predictions about how the sepia eye color trait will be passed down from one generation to the next. If you can use your model to make accurate predictions about how the traits of the flies are inherited, then you will be able to generate the evidence you need to convince others that the conceptual model you decided to use is the most appropriate one.

Connections to Crosscutting Concepts, the Nature of Science, and the Nature of Scientific Inquiry

As you work through your investigation, be sure to think about

- the importance of uncovering causes for patterns observed in nature,
- how scientists develop and use explanatory models to make sense of their observations,
- the nature of theories and laws in science, and
- the difference between data and evidence in science.

Initial Argument

Once your group has finished collecting and analyzing your data, you will need to develop an initial argument. Your argument must include a claim, evidence to support your claim, and a justification of the evidence. The claim is your group's answer to the guiding question. The evidence is an analysis and interpretation of your data. Finally, the justification of the evidence is why your group thinks the evidence matters. The justification of the evidence is important because scientists can use different kinds of evidence to support their claims. Your group will create your initial argument on a whiteboard. Your whiteboard should include all the information shown in Figure L16.2.

FIGURE L16.2

Argument presentation on a whiteboard

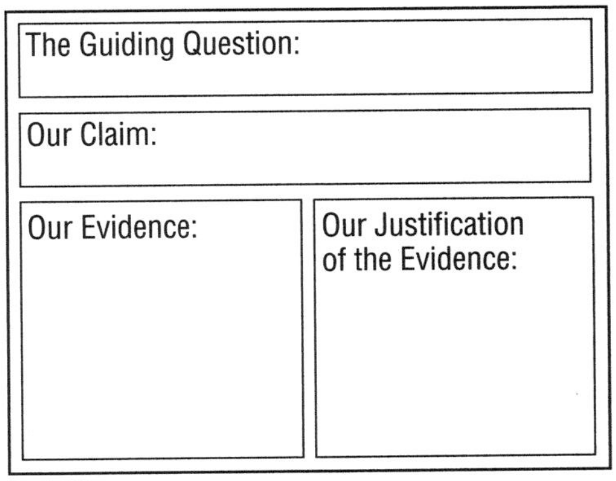

The Guiding Question:	
Our Claim:	
Our Evidence:	Our Justification of the Evidence:

Argumentation Session

The argumentation session allows all of the groups to share their arguments. One member of each group will stay at the lab station to share that group's argument, while the other members of the group go to the other lab stations one at a time to listen to and critique the arguments developed by their classmates. This is similar to how scientists present their arguments to other scientists at conferences. If you are responsible for critiquing your classmates' arguments, your goal is to look for mistakes so these mistakes can be fixed and they can make their argument better. The argumentation session is also a good time to think about ways you can make your initial argument better. Scientists must share and critique arguments like this to develop new ideas.

LAB 16

To critique an argument, you might need more information than what is included on the whiteboard. You will therefore need to ask the presenter lots of questions. Here are some good questions to ask:

- What did your group do to collect the data? Why do you think that way is the best way to do it?
- What did your group do to analyze the data? Why did your group decide to analyze it that way?
- What other ways of analyzing and interpreting the data did your group talk about?
- What did your group do to make sure that these calculations are correct?
- Why did your group decide to present your evidence in that way?
- What other claims did your group discuss before you decided on that one? Why did your group abandon those other ideas?
- How sure are you that your group's claim is accurate? What could you do to be more certain?

Once the argumentation session is complete, you will have a chance to meet with your group and revise your initial argument. Your group might need to gather more data as part of this process. Remember, your goal at this stage of the investigation is to develop the best argument possible.

Report

Once you have completed your research, you will need to prepare an investigation report that consists of three sections that provide answers to the following questions:

1. What question were you trying to answer and why?
2. What did you do during your investigation and why did you conduct your investigation in this way?
3. What is your argument?

Your report should answer these questions in two pages or less. This report must be typed and any diagrams, figures, or tables should be embedded into the document. Be sure to write in a persuasive style; you are trying to convince others that your claim is acceptable or valid!

Checkout Questions

Lab 16. Mechanisms of Inheritance: How Do Fruit Flies Inherit the Sepia Eye Color Trait?

1. In your own words, explain the difference between the dominant-recessive model of inheritance and the incomplete dominance model of inheritance.

2. Explain how a certain type of flower has two types of alleles for color, such as red and white, but produces red, white, and pink flowers.

3. Laws are theories that have been proven true.

   a. I agree with this statement.

   b. I disagree with this statement.

   Explain your answer, using an example from your investigation about mechanisms of inheritance.

4. Data are observations or measurements collected during an investigation, and evidence is analyzed data and an interpretation of the analysis.

   a. I agree with this statement.
   b. I disagree with this statement.

   Explain your answer, using an example from your investigation about mechanisms of inheritance.

5. Scientists often look for patterns when they study nature. Explain why the identification of patterns in nature is so important, using an example from your investigation about mechanisms of inheritance.

6. Scientists often need to use or create conceptual models to understand a natural phenomenon. Discuss why models are important in science, using an example from your investigation about mechanisms of inheritance.

SECTION 5

Life Sciences
Core Idea 4

Biological Evolution: Unity and Diversity

Introduction Lab

LAB 17

Lab 17. Mechanisms of Evolution: Why Does a Specific Version of a Trait Become More Common in a Population Over Time?

Purpose

The purpose of this lab is to *introduce* students to the basic principles of natural selection. Specifically, this investigation gives students an opportunity to use a simulation to develop an explanation for a natural phenomenon. The simulation, called *Bug Hunt Camouflage* (Novak and Wilensky 2005), was created using NetLogo, a multiagent programmable modeling environment developed at the Center for Connected Learning and Computer-Based Modeling at Northwestern University (Wilensky 1999).

This lab gives students an opportunity to learn how scientists use models to understand complex phenomena and to study cause-and-effect relationships. Students will also have the opportunity to reflect on the different types of methods that scientists use in order to answer questions and the important role that imagination and creativity play in science.

The Content

Natural selection is a mechanism that drives changes in the characteristics of a population. The basic tenets of natural selection are as follows (Lawson 1995):

- Only a fraction of the individuals that make up a population survive long enough to reproduce.

- The individuals in a population are not all the same. Individuals have traits or versions of a trait that make them unique.

- Much, but not all, of this variation in traits is inheritable and can therefore be passed down from parent to offspring.

- The environment, including both abiotic (e.g., temperature, amount of water available) and biotic (e.g., amount of food, presence of predators) factors, determines which traits are favorable or unfavorable, because some traits increase an individual's chance of survival and others do not.

- Individuals with a favorable trait or a version of a trait tend to produce more offspring than those with unfavorable ones. Therefore, over time, a favorable trait or a version of a trait will become more common within a population found in a particular environment (and unfavorable traits become less common).

To learn more about natural selection and the role it plays in evolution, we recommend visiting the Understanding Evolution website at *http://evolution.berkeley.edu/evolibrary/home.php*. This website was created by the University of California Museum of Paleontology and is an excellent resource for teachers and students.

Timeline

The instructional time needed to implement this lab investigation is 130–200 minutes. Appendix 2 (p. 355) provides options for implementing this lab investigation over several class periods. Option C (200 minutes) should be used if students are unfamiliar with scientific writing, because this option provides extra instructional time for scaffolding the writing process. You can scaffold the writing process by modeling, providing examples, and providing hints as students write each section of the report. Option D (130 minutes) should be used if students are familiar with scientific writing and have the skills needed to write an investigation report on their own. In option D, students complete stage 6 (writing the investigation report) and stage 8 (revising the investigation report) as homework.

Materials and Preparation

The materials needed to implement this investigation are listed in Table 17.1. The *Bug Hunt Camouflage* simulation, available at *http://ccl.northwestern.edu/netlogo/models/BugHuntCamouflage,* is free to use and can be downloaded as part of an application. You should access the simulation and learn how it works before beginning the lab investigation. In addition, it is important to check if students can access and use the simulation from a school computer, because some schools have set up firewalls and other restrictions on web browsing.

Since this is an introduction lab, students do not have any formal instruction about evolution or the mechanisms of evolution before the lab.

TABLE 17.1

Materials list

Item	Quantity
Computer with NetLogo application	At least 1 per group
Investigation Proposal A	1 per group
Whiteboard, 2' × 3'*	1 per group
Lab Handout	1 per student
Peer-review guide	1 per student
Checkout Questions	1 per student

* Students can also use computer and presentation software such as Microsoft PowerPoint or Apple Keynote to create their arguments.

LAB 17

Safety Precautions

Follow all normal lab safety rules.

Topics for the Explicit and Reflective Discussion

Concepts That Can Be Used to Justify the Evidence

To provide an adequate justification of their evidence, students must explain why they included the evidence in their arguments and make the assumptions underlying their analysis and interpretation of the data explicit. In this investigation, students can use the following concepts to help justify their evidence:

- Population dynamics
- Predation and predator-prey relationships
- Genetic basis of inheritance
- Natural selection

We recommend that you review these concepts during the explicit and reflective discussion to help students make this connection.

How to Design Better Investigations

It is important for students to reflect on the strengths and weaknesses of the investigation they designed during the explicit and reflective discussion. Students should therefore be encouraged to discuss ways to eliminate potential flaws, measurement errors, or sources of bias in their investigations. To help students be more reflective about the design of their investigation, you can ask the following questions:

- What were some of the strengths of your investigation? What made it scientific?
- What were some of the weaknesses of your investigation? What made it less scientific?
- If you were to do this investigation again, what would you do to address the weaknesses in your investigation? What could you do to make it more scientific?

Crosscutting Concepts

This investigation is aligned with two crosscutting concepts found in *A Framework for K–12 Science Education*, and you should review these concepts during the explicit and reflective discussion.

- *Cause and effect: Mechanism and explanation:* An important goal of science is to uncover the underlying cause of a natural phenomenon.

- *Systems and system models:* It is critical for scientists to be able to define the system under study (e.g., the components of a habitat) and then make a model of it to understand it. Models can be physical, conceptual, or mathematical.

The Nature of Science and the Nature of Scientific Inquiry

This investigation is aligned with two important concepts related to the *nature of science* (NOS) and the *nature of scientific inquiry* (NOSI), and you should review these concepts during the explicit and reflective discussion.

- *Methods used in scientific investigations:* Examples of methods include experiments, systematic observations of a phenomenon, literature reviews, and analysis of existing data sets; the choice of method depends on the objectives of the research. There is no universal step-by-step scientific method that all scientists follow; rather, different scientific disciplines (e.g., biology vs. physics) and fields within a discipline (e.g., ecology vs. molecular biology) use different types of methods, use different core theories, and rely on different standards to develop scientific knowledge. In this investigation, for example, students use a simulation; they do not conduct a field study.

- *The importance of imagination and creativity in science:* Students should learn that developing explanations for or models of natural phenomena and then figuring out how they can be put to the test of reality is as creative as writing poetry, composing music, or designing skyscrapers. Scientists must also use their imagination and creativity to figure out new ways to test ideas and collect or analyze data.

Hints for Implementing the Lab

- A group of three students per computer tends to work well.

- Allow the students to play with the simulation as part of the tool talk before they fill out an investigation proposal. This gives students a chance to see what they can and cannot do with the simulation.

- Be sure that students record actual values (e.g., number of bugs with specific traits at a given point in time) when they run a simulation, rather than just attempting to hand draw the graph that they see on the computer screen.

- This investigation provides a great introduction to natural selection. We recommend that you explain this important concept after this lab is complete and then use this investigation to illustrate each of the major tenets of natural selection.

Topic Connections

Table 17.2 (p. 284) provides an overview of the scientific practices, crosscutting concepts, disciplinary core ideas, and support ideas at the heart of this lab investigation. In addition, it lists NOS and NOSI concepts for the explicit and reflective discussion. Finally, it lists

LAB 17

literacy and mathematics skills (*CCSS ELA* and *CCSS Mathematics*) that are addressed during the investigation.

TABLE 17.2

Lab 17 alignment with standards

Scientific practices	• Asking questions and defining problems • Developing and using models • Planning and carrying out investigations • Analyzing and interpreting data • Using mathematics and computational thinking • Constructing explanations • Engaging in argument from evidence • Obtaining, evaluating, and communicating information
Crosscutting concepts	• Patterns • Cause and effect: Mechanism and explanation • Systems and system models
Core ideas	• LS2: Ecosystems: Interactions, energy, and dynamics • LS3: Heredity: Inheritance and variation of traits • LS4: Biological evolution: Unity and diversity
Supporting ideas	• Population dynamics • Predation • Predators • Prey • Inheritance of traits • Natural selection
NOS and NOSI concepts	• Methods used in scientific investigations • Imagination and creativity in science
Literacy connections (*CCSS ELA*)	• *Reading:* Key ideas and details, craft and structure, integration of knowledge and ideas • *Writing:* Text types and purposes, production and distribution of writing, research to build and present knowledge, range of writing • *Speaking and listening:* Comprehension and collaboration, presentation of knowledge and ideas
Mathematics connections (*CCSS Mathematics*)	• Make sense of problems and persevere in solving them • Reason abstractly and quantitatively • Construct viable arguments and critique the reasoning of others • Use appropriate tools strategically

References

Lawson, A. 1995. *Science teaching and the development of thinking.* Belmont, CA: Wadsworth.

Novak, M., and U. Wilensky. 2005. NetLogo Bug Hunt Camouflage model. Evanston, IL: Center for Connected Learning and Computer-Based Modeling, Northwestern Institute on Complex Systems, Northwestern University. *http://ccl.northwestern.edu/netlogo/models/BugHuntCamouflage.*

Wilensky, U. 1999. NetLogo. Evanston, IL: Center for Connected Learning and Computer-Based Modeling, Northwestern Institute on Complex Systems, Northwestern University. *http://ccl.northwestern.edu/netlogo.*

LAB 17

Lab Handout

Lab 17. Mechanisms of Evolution: Why Does a Specific Version of a Trait Become More Common in a Population Over Time?

Introduction

An *ecosystem* includes all the organisms and the nonliving parts of the environment that are found in a particular area. Organisms include things such as plants, animals, fungi, and bacteria. The nonliving parts of the environment include things such as air, light, water, and minerals. The organisms found within an ecosystem depend on the nonliving components for survival. The organisms also interact with each other. For example, plants need air, light, and water to produce the food they need to survive. Animals called herbivores eat these plants. Other animals called predators eat the herbivores. Herbivores and predators also need water to drink and air to breathe in order to survive. All the living and nonliving parts of the environment therefore function as a system. A change in one part of the system will, as a result, affect the other parts of the system. For example, a drought could reduce the number of plants in a particular area. A decrease in the number of plants will result in less food for the herbivores. When these animals do not have enough food to eat, some will starve. The predators will then not have enough food to eat.

Organisms often have adaptations that allow them to function in a specific ecosystem. An adaptation can be a physical feature that helps an organism to survive. Katydids, for example, are insects that look like leaves (Figure L17.1), and their unique appearance helps them to avoid predators. An adaptation can also be something that an organism is able to do that helps it survive in a specific environment. The creosote bush (Figure L17.2), for example, reduces competition for nutrients and water by producing a toxin that prevents other plants from growing near it. Biologists define an *adaptation* as a version of a trait that is common in a population because it provides some improved function over other versions of that trait.

Organisms that live in different ecosystems tend to have different adaptations. For example, a population of herbivores that lives in an ecosystem with a lot of predators will have different adaptations than

FIGURE L17.1

A katydid

Katydids are insects that look like a leaf. The appearance of the insect is an adaptation.

FIGURE L17.2

A creosote bush

The creosote bush is a desert-dwelling plant that produces a toxin that prevents the growth of other plants.

a population of herbivores that lives in an ecosystem with few or no predators. Similarly, a herbivore population that lives in an ecosystem that gets very little rain will have different adaptations than a herbivore population that lives in an ecosystem that gets a lot of rain. It is therefore important for biologists to understand why specific versions of a trait become more or less common in a population and how changes in an ecosystem will affect the characteristics of the organisms in it. In this investigation, you will examine how a specific trait in a simulated population of bugs changes over time in two different environments. You will then develop a conceptual model that can be used to explain why a certain version of a trait becomes more common in a population over several generations.

Your Task

Use a computer simulation to explore how the frequency of different versions of the body color trait in a population of bugs changes over time in two different environments. You will then develop a model that can be used to explain why a certain version of a trait becomes more common in a population over several generations.

The guiding question of this investigation is, **Why does a specific version of a trait become more common in a population over time?**

Materials

You will use an online simulation called *Bug Hunt Camouflage* to conduct your investigation. You can find the simulation by going to the following website: *http://ccl.northwestern. edu/netlogo/models/BugHuntCamouflage*.

Safety Precautions

Follow all normal lab safety rules.

Investigation Proposal Required?   ☐ Yes   ☐ No

Getting Started

The first step in developing your model will be to use the *Bug Hunt Camouflage* simulation to explore how the frequency of different versions of the body color trait in a simulated population of bugs changes over time in two different environments. The bugs that make up the simulated population belong to the same species but are different colors. There are, as a result, individual bugs with different versions of the body color trait in the simulated environment. The color of each bug in this simulation is described in terms of its hue, saturation, and brightness (HSB). Hue is a color such as red, green, or blue and is given a value ranging from 0 to 255 in this simulation. Saturation is the purity or richness of a color and ranges from 0 (gray) to 255 (colorful) in the simulation. Brightness is the intensity of the color and, like saturation, ranges from 0 (dark) to 255 (bright) in the simulation.

Remember, all of the bugs in the simulated environment are from the same species, even though they look different.

In this simulation, you will act as the predator. You can eat the bugs (your prey) by clicking on them. When a bug is eaten, it is replaced through reproduction by another bug in the simulated ecosystem. The new bug will often (but not always) have the same color as the parent bug. The simulation provides information about the total number of bugs that you have caught since the simulation started, the current color composition of the bug population (in terms of HSB), and how the average color of the bugs in the population has changed over time. You can also change the number of bugs that are in the environment at any given time (carrying capacity) and the environment type (glacier, beach, poppy field). There are several other factors, such as bug size, that you can adjust as part of the simulation. Figure L17.3 illustrates the factors in the simulation.

FIGURE L17.3

A screen shot from the *Bug Hunt Camouflage* simulation

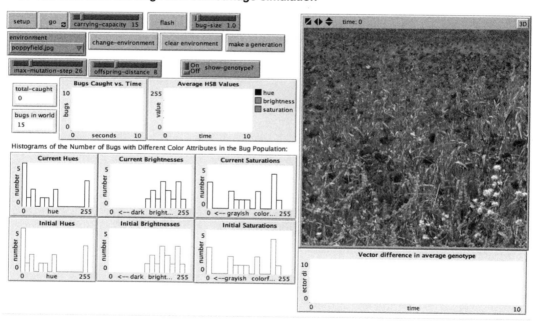

To explore how the frequency of different versions of the body color trait in the population of bugs changes over time in two different environments, you must determine what type of data you need to collect and how you will collect it using the *Bug Hunt Camouflage* simulation. You also need to determine how you will analyze the data once it has been collected. To determine *what type of data you need to collect*, think about the following questions:

- What will you do to track how the body color trait in the bug population changes over time?
- What will serve as your dependent variable (e.g., average HSB value, current hue, current brightness, number of bugs caught)?
- What type of measurements or observations will you need to record during your investigation?

To determine *how you will collect your data*, think about the following questions:

- What will serve as a control condition?
- What types of treatment conditions will you need to set up and how will you do it?
- How many trials will you need to conduct?
- How long will you need to run the simulation during each trial (e.g., for three minutes or until 60 bugs are caught)?
- How often will you record an observation or collect a measurement?
- When will you make your observations or make a measurement?
- How will you keep track of the data you collect and how will you organize it?

To determine *how you will analyze your data*, think about the following questions:

- How will you determine if there is a difference between the different treatment conditions and the control condition?
- What type of calculations will you need to make?
- What type of graph could you create to help make sense of your data?

Once you have collected and analyzed your data, your group will need to develop a conceptual model to explain why a specific version of the body color trait becomes more common in the population over several generations. Your model, however, should also be able to explain how traits in other populations of organisms can change over time. It will therefore be important for you to think about how your model could be used to explain a wide range of situations and not just what happened in your investigation.

The last step in this investigation is to test your model. To accomplish this goal, you can use a third environment in the *Bug Hunt Camouflage* simulation to determine if you can use your model to make accurate predictions about how the bug color trait changes over several generations under different conditions. If you can use your model to make accurate predictions about how the traits of the bugs in the population change in a new environment, then you will be able to generate the evidence you need to convince others that the conceptual model you developed is valid.

LAB 17

Connections to Crosscutting Concepts, the Nature of Science, and the Nature of Scientific Inquiry

As you work through your investigation, be sure to think about

- the importance of looking for patterns in nature,
- the importance of developing explanations for a natural phenomenon,
- how scientists create and use models to understand a natural phenomenon,
- the different types of methods that scientists use to answer questions, and
- the important role that imagination and creativity play in science.

Initial Argument

Once your group has finished collecting and analyzing your data, you will need to develop an initial argument. Your argument must include a claim, evidence to support your claim, and a justification of the evidence. The claim is your group's answer to the guiding question. The evidence is an analysis and interpretation of your data. Finally, the justification of the evidence is why your group thinks the evidence matters. The justification of the evidence is important because scientists can use different kinds of evidence to support their claims. Your group will create your initial argument on a whiteboard. Your whiteboard should include all the information shown in Figure L17.4.

FIGURE L17.4

Argument presentation on a whiteboard

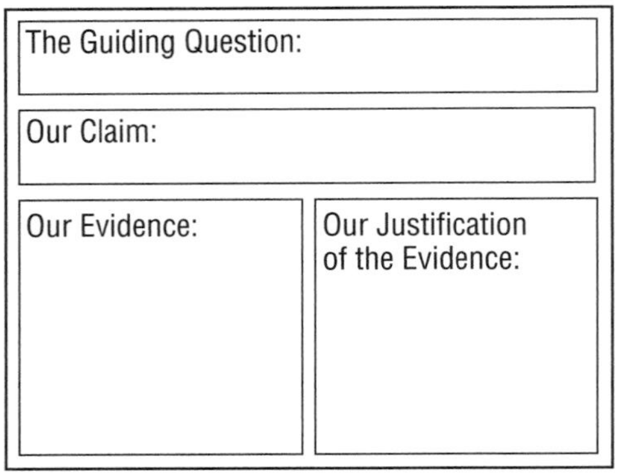

The Guiding Question:	
Our Claim:	
Our Evidence:	Our Justification of the Evidence:

Argumentation Session

The argumentation session allows all of the groups to share their arguments. One member of each group will stay at the lab station to share that group's argument, while the other members of the group go to the other lab stations one at a time to listen to and critique the arguments developed by their classmates. This is similar to how scientists present their arguments to other scientists at conferences. If you are responsible for critiquing your classmates' arguments, your goal is to look for mistakes so these mistakes can be fixed and they can make their argument better. The argumentation session is also a good time to think about ways you can make your initial argument better. Scientists must share and critique arguments like this to develop new ideas.

To critique an argument, you might need more information than what is included on the whiteboard. You will therefore need to ask the presenter lots of questions. Here are some good questions to ask:

- What did your group do to collect the data? Why do you think that way is the best way to do it?

- What did your group do to analyze the data? Why did your group decide to analyze it that way?
- What other ways of analyzing and interpreting the data did your group talk about?
- What did your group do to make sure that these calculations are correct?
- Why did your group decide to present your evidence in that way?
- What other claims did your group discuss before you decided on that one? Why did your group abandon those other ideas?
- How sure are you that your group's claim is accurate? What could you do to be more certain?

Once the argumentation session is complete, you will have a chance to meet with your group and revise your initial argument. Your group might need to gather more data as part of this process. Remember, your goal at this stage of the investigation is to develop the best argument possible.

Report

Once you have completed your research, you will need to prepare an investigation report that consists of three sections that provide answers to the following questions:

1. What question were you trying to answer and why?
2. What did you do during your investigation and why did you conduct your investigation in this way?
3. What is your argument?

Your report should answer these questions in two pages or less. The report must be typed, and any diagrams, figures, or tables should be embedded into the document. Be sure to write in a persuasive style; you are trying to convince others that your claim is acceptable or valid!

Checkout Questions

Lab 17. Mechanisms of Evolution: Why Does a Specific Version of a Trait Become More Common in a Population Over Time?

Snowshoe hares live in the boreal forests of Alaska, Washington, Idaho, Montana, and Canada. In winter, they grow long white guard hairs that match the snow (see the figure on the left, below). In summer, they shed their white guard hairs and have mostly rusty brown coats that blend in with trees and soil (see the figure on the right, below). Snowshoe hares are able to hide from predators (including lynx, coyotes, foxes, wolves, and birds of prey) because they are able to blend into their surroundings.

A snowshoe hare with white fur

A snowshoe hare with brown fur

The signal for a hare to shift coat color comes from the pineal gland in the brain, which senses changes in daylight length. When the days of fall get shorter, it triggers the coat color to change from brown to white. When the days get longer in the spring, the white hairs begin to shed. Usually, shorter days correspond with colder temperatures and more snowfall, so the snowshoe hare is usually white when the ground is covered with snow.

Unfortunately, the average temperature in Alaska, Washington, Idaho, Montana, and Canada has increased over the last decade and the ground is not covered in snow until well into the winter. The snowshoe hare, however, still changes color regardless of when there is snow on the ground because the shift in coat color is triggered by daylight length rather than temperature. As a result, many snowshoe hares turn white before it snows and these white hares tend to stand out against the brown background of trees and soil. Biologists have observed that the population of snowshoe hares found in these geographic areas is getting smaller because predators are catching more and more hares each fall. However, biologists also predict that the snowshoe hare population will adapt to this change in the environment.

1. Use what you have learned about how populations evolve over time to explain how this snowshoe hare population could adapt to warmer temperatures.

2. All scientists use the same method to test their ideas.

   a. I agree with this statement.

   b. I disagree with this statement.

   Explain your answer, using an example from your investigation about the mechanisms of evolution.

3. Scientists do not need to be creative or have a good imagination to be successful in science.

   a. I agree with this statement.

   b. I disagree with this statement.

   Explain your answer, using an example from your investigation about the mechanisms of evolution.

4. Scientists often attempt to identify patterns in nature. Explain why the identification of patterns is useful in science, using an example from your investigation about the mechanisms of evolution.

5. An important goal in science is to develop explanations for natural phenomena. Explain why the development of explanations is so important in science, using an example from your investigation about the mechanisms of evolution.

6. Scientists often attempt to develop models of systems in order to study them. Explain why developing a model of a system is useful in science, using an example from your investigation about the mechanisms of evolution.

Application Labs

Teacher Notes

Lab 18. Environmental Change and Evolution: Which Mechanism of Microevolution Caused the Beak of the Medium Ground Finch Population on Daphne Major to Increase in Size From 1976 to 1978?

Purpose

The purpose of this lab is for students to *apply* what they know about migration, genetic drift, and natural selection to explain the evolution of beak size in a population of birds. Specifically, this investigation gives students an opportunity to use an existing data set to test three different potential explanations for a case of microevolution. Through this activity, students will have an opportunity to learn how scientists use system models to understand natural phenomena and to learn about the connection between structure and function in living things. Students will also have the opportunity to reflect on the difference between theories and laws in science and on the various methods that scientists can use during an investigation.

The Content

Microevolution is a change in gene frequency in a population over time. A population is a group of organisms that share a common gene pool, and a population of animals is a group of individuals that live in the same area and are able to mate and produce fertile offspring. Figure 18.1 provides an illustration of microevolution in a population of mosquitoes. In this example, a gene for pesticide resistance becomes more common in the mosquito population over time.

There are four basic mechanisms of microevolution: *mutation, migration, natural selection,* and *genetic drift.* Any one these four mechanisms can affect the frequency of a gene in a population. These four mechanisms can also work in combination.

The first mechanism of microevolution is a genetic mutation. A mutation during the DNA replication process can result in an individual being born with a new version of a gene. The individual with the new gene can then have offspring with the same gene. The new gene could then become more common in a population over time. Figure 18.2 provides an illustration of how a genetic mutation can lead to change in the frequency of a gene for pesticide resistance within a population of mosquitoes over time. It is important to note, however, that mutations are rare and only happen in individuals. Genetic mutations therefore cannot result in a big change in the frequency of a gene within a population in only one or two generations.

Environmental Change and Evolution

Which Mechanism of Microevolution Caused the Beak of the Medium Ground Finch Population on Daphne Major to Increase in Size From 1976 to 1978?

FIGURE 18.1

Microevolution in a population of mosquitoes

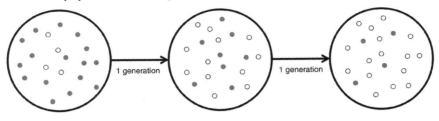

20% of the mosquitos carry a gene for pesticide resistance

60% of the mosquitos carry a gene for pesticide resistance

80% of the mosquitos carry a gene for pesticide resistance

Note: The white dots within each circle represent the gene for pesticide resistance.

FIGURE 18.2

Microevolution in a population of mosquitoes due to a genetic mutation

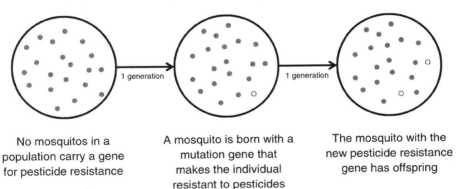

No mosquitos in a population carry a gene for pesticide resistance

A mosquito is born with a mutation gene that makes the individual resistant to pesticides

The mosquito with the new pesticide resistance gene has offspring

Note: The white dots within each circle represent the gene for pesticide resistance.

The second mechanism of microevolution is migration, which is also known as *gene flow*. Individuals can either join a population (immigration) or leave a population (emigration). A specific version of gene will become less common within a population when several individuals with that gene leave the population, and a specific version of a gene will become more common within a population when several individuals with that gene join the population. The migration of a large number of individuals into or out of a population can therefore result in a dramatic shift in the frequency of a gene within a population in a relatively short period of time. Figure 18.3 (p. 298) provides an illustration of how migration or gene flow can change the frequency of a gene for pesticide resistance in a population of mosquitoes over time.

LAB 18

FIGURE 18.3

Microevolution in a population of mosquitoes due to migration (gene flow)

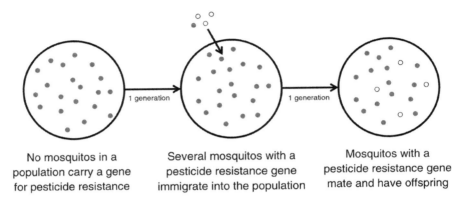

| No mosquitos in a population carry a gene for pesticide resistance | Several mosquitos with a pesticide resistance gene immigrate into the population | Mosquitos with a pesticide resistance gene mate and have offspring |

Note: The white dots within each circle represent the gene for pesticide resistance.

Natural selection is the third mechanism of microevolution. Natural selection occurs when

- there is variation in a trait among the individuals that make up a population,
- the trait is determined by one or more genes,
- the trait affects survival and/or ability to reproduce, and
- individuals who reproduce pass on their genes to the next generation.

The frequency of a gene in any given generation, as a result, reflects the traits and genes of the individuals that were able to survive long enough to reproduce in the previous generation. Over time, genes that determine traits that are associated with an increased chance of survival and successful reproduction will become more common in a population, and the genes that determine traits that decrease an individual's chance of survival or reproduction will become less common. Figure 18.4 provides an illustration of how natural selection can change the frequency of the gene for pesticide resistance in a population of mosquitoes over time.

The fourth mechanism of microevolution is genetic drift. In any generation, some individuals may, just by chance, survive longer or leave behind more offspring than other individuals. The frequency of a gene in the next generation will therefore reflect the genes and traits of these lucky individuals rather than individuals with traits that are advantageous in terms of survival or reproduction. This process causes the frequency of genes in a population to change (or drift) over time. Genetic drift tends to act faster and has more drastic results in smaller populations. It also tends to decrease genetic variation in populations.

Environmental Change and Evolution

Which Mechanism of Microevolution Caused the Beak of the Medium Ground Finch Population on Daphne Major to Increase in Size From 1976 to 1978?

FIGURE 18.4

Microevolution in a population of mosquitoes due to natural selection

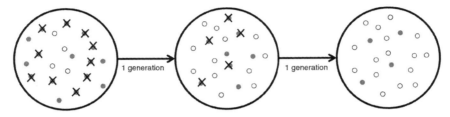

Some mosquitos in a population carry a gene for pesticide resistance and some do not. Individuals with the pesticide resistance are more likely to survive and reproduce because people spray with pesticides. The gene for pesticide resistance becomes more common in the population of each generation.

Note: The white dots within each circle represent the gene for pesticide resistance, and the Xs represent individual mosquitoes that do not survive.

In this lab, the students study a specific bird population, the medium ground finch (*Geospiza fortis*), that lives in the Galápagos Islands, an archipelago of volcanic islands in the Pacific Ocean. The major factor that affects the survival of these birds is the availability of food. The medium ground finch has a small beak and prefers to eat small seeds with soft shells. In 1977, the islands received very little rain. The plants on the island withered and stopped producing seeds. The medium ground finches quickly depleted the supply of small seeds with soft shells. There were, however, still large seeds with hard shells on the island. The finches with larger beaks were able to crack open and eat the larger seeds, but the smaller-beaked birds were not. The smaller-beaked birds therefore died of starvation and the larger-beaked birds survived the drought and reproduced. The average beak size in the next generation of these finches, as a result, was about 1 mm larger than the previous generation.

Timeline

The instructional time needed to implement this lab investigation is 130–200 minutes. Appendix 2 (p. 355) provides options for implementing this lab investigation over several class periods. Option C (200 minutes) should be used if students are unfamiliar with scientific writing, because this option provides extra instructional time for scaffolding the writing process. You can scaffold the writing process by modeling, providing examples, and providing hints as students write each section of the report. Option D (130 minutes) should be used if students are familiar with scientific writing and have the skills needed to write an investigation report on their own. In option D, students complete stage 6 (writing the investigation report) and stage 8 (revising the investigation report) as homework.

LAB 18

Materials and Preparation

The materials needed to implement this investigation are listed in Table 18.1. The *Finch Data.xls* file is available at *www.nsta.org/publications/press/extras/adi-lifescience.aspx*. You should download the file and explore it before beginning the lab investigation. Since this is an "application" lab, students should be familiar with the mechanisms of microevolution before the lab.

The data found in the Finch Data Excel files are based on the published work of Peter R. Grant, B. Rosemary Grant, and their colleagues who have studied the Medium Ground Finches on Daphne Major for the past four decades. The actual data are drawn from the following sources: Boag and Grant 1981, 1984; Grant 1989; Grant and Grant 1980, 2002. The individual bird characteristics, amount of rainfall, and seed type abundance data provide a simplified data set, consistent in all respects with the published data but with fewer data points, to make these data more accessible to students.

TABLE 18.1 _____

Materials list

Item	Quantity
Computer with a spreadsheet application such as Microsoft Excel or Apple Numbers	At least 1 per group
Finch Data Excel file	At least 1 per group
Investigation Proposal B (optional)*	1 per group
Whiteboard, 2' × 3'†	1 per group
Lab Handout	1 per student
Peer-review guide	1 per student
Checkout Questions	1 per student

* We highly recommend that students fill out an investigation proposal for this lab.

† Students can also use computer and presentation software such as Microsoft PowerPoint or Apple Keynote to create their arguments.

Safety Precautions

Follow all normal lab safety rules.

Topics for the Explicit and Reflective Discussion

Concepts That Can Be Used to Justify the Evidence

To provide an adequate justification of their evidence, students must explain why they included the evidence in their arguments and make the assumptions underlying their analysis and interpretation of the data explicit. In this investigation, students can use the following concepts to help justify their evidence:

- Population dynamics
- Inheritance of traits
- Microevolution
- Mutation
- Migration
- Natural selection
- Genetic drift

We recommend that you review these concepts during the explicit and reflective discussion to help students make this connection.

How to Design Better Investigations

It is important for students to reflect on the strengths and weaknesses of the investigation they designed during the explicit and reflective discussion. Students should therefore be encouraged to discuss ways to eliminate potential flaws, measurement errors, or sources of bias in their investigations. To help students be more reflective about the design of their investigation, you can ask the following questions:

- What were some of the strengths of your investigation? What made it scientific?
- What were some of the weaknesses of your investigation? What made it less scientific?
- If you were to do this investigation again, what would you do to address the weaknesses in your investigation? What could you do to make it more scientific?

Crosscutting Concepts

This investigation is aligned with two crosscutting concepts found in *A Framework for K–12 Science Education,* and you should review these concepts during the explicit and reflective discussion.

- *Systems and system models:* Scientists often need to define the system they are studying (e.g., the components of a habitat) and then use a model to understand it. Models can be physical, conceptual, or mathematical.
- *Structure and function:* In nature, the way a living thing is shaped or structured determines how it functions and places limits on what it can and cannot do. In this investigation, for example, beak shape affected a bird's ability to eat.

LAB 18

The Nature of Science and the Nature of Scientific Inquiry

This investigation is aligned with two important concepts related to the *nature of science* (NOS) and the *nature of scientific inquiry* (NOSI), and you should review these concepts during the explicit and reflective discussion.

- *The difference between laws and theories in science:* A scientific law describes the behavior of a natural phenomenon or a generalized relationship under certain conditions; a scientific theory is a well-substantiated explanation of some aspect of the natural world. Theories do not become laws even with additional evidence; they explain laws. However, not all scientific laws have an accompanying explanatory theory. It is also important for students to understand that scientists do not discover laws or theories; the scientific community develops them over time.

- *Methods used in scientific investigations:* Examples of methods include experiments, systematic observations of a phenomenon, literature reviews, and analysis of existing data sets; the choice of method depends on the objectives of the research. There is no universal step-by step scientific method that all scientists follow; rather, different scientific disciplines (e.g., biology vs. physics) and fields within a discipline (e.g., ecology vs. molecular biology) use different types of methods, use different core theories, and rely on different standards to develop scientific knowledge. In this investigation, for example, students use a large data set; they do not conduct a field study.

Hints for Implementing the Lab

- Show students how to use the spreadsheet application as part of the tool talk. At a minimum, students will need to know how to use formulas, make new sheets, and create charts.

- The students should be encouraged to think of ways to use the available data to test the three potential explanations for the evolutionary change in beak size. We recommend that students fill out Investigation Proposal B at the beginning of the lab to help them generate predictions based on each explanation.

- Encourage students to make a copy of the Finch Data Excel file using the "Save as" feature before they start analyzing the data.

- Students can cut and paste the data into new sheets to facilitate analysis.

- A group of three students per computer tends to work well.

- Have students create charts in the spreadsheet application for the argumentation sessions and investigation reports.

- Students may not be able to refute one or more alternative explanations due to limitations in the data set. Be sure to remind students to acknowledge the limitations in the data and then encourage them to think about what other data they would need to determine which explanation is the most valid or acceptable.

Environmental Change and Evolution

Which Mechanism of Microevolution Caused the Beak of the Medium Ground Finch Population on Daphne Major to Increase in Size From 1976 to 1978?

Topic Connections

Table 18.2 provides an overview of the scientific practices, crosscutting concepts, disciplinary core ideas, and support ideas at the heart of this lab investigation. In addition, it lists NOS and NOSI concepts for the explicit and reflective discussion. Finally, it lists literacy and mathematics skills (*CCSS ELA* and *CCSS Mathematics*) that are addressed during the investigation.

TABLE 18.2

Lab 18 alignment with standards

Scientific practices	• Asking questions and defining problems • Developing and using models • Planning and carrying out investigations • Analyzing and interpreting data • Using mathematics and computational thinking • Constructing explanations • Engaging in argument from evidence • Obtaining, evaluating, and communicating information
Crosscutting concepts	• Systems and system models • Structure and function
Core ideas	• LS2: Ecosystems: Interactions, energy, and dynamics • LS3: Heredity: Inheritance and variation of traits • LS4: Biological evolution: Unity and diversity
Supporting ideas	• Population dynamics • Inheritance of traits • Microevolution • Mutation • Migration • Natural selection • Genetic drift
NOS and NOSI concepts	• Scientific laws and theories • Methods used in scientific investigations
Literacy connections (*CCSS ELA*)	• *Reading:* Key ideas and details, craft and structure, integration of knowledge and ideas • *Writing:* Text types and purposes, production and distribution of writing, research to build and present knowledge, range of writing • *Speaking and listening:* Comprehension and collaboration, presentation of knowledge and ideas
Mathematics connections (*CCSS Mathematics*)	• Make sense of problems and persevere in solving them • Reason abstractly and quantitatively • Construct viable arguments and critique the reasoning of others • Model with mathematics • Use appropriate tools strategically • Look for and express regularity in repeated reasoning

LAB 18

References

Boag, P. T., and P. R. Grant. 1981. Intense natural selection in a population of Darwin's finches (Geospizinae) in the Galápagos. *Science* 214: 82–85.

Boag, P. T., and P. R. Grant. 1984. Darwin's finches (Geospiza) on Isla Daphne Major, Galápagos: Breeding and feeding ecology in a climatically variable environment. *Ecological Monographs* 54: 463–489.

Grant, P. R. 1989. *Ecology and evolution of Darwin's finches.* Princeton, NJ: Princeton University Press.

Grant, P. R., and B. R. Grant. 1980. Annual variation in finch numbers, foraging and food supply on Isla Daphne Major, Galápagos. *Oecologia* 46: 55–62.

Grant, P. R., and B. R. Grant. 2002. Unpredictable evolution in a 30-year study of Darwin's finches. *Science* 296: 707–711.

Lab Handout

Lab 18. Environmental Change and Evolution: Which Mechanism of Microevolution Caused the Beak of the Medium Ground Finch Population on Daphne Major to Increase in Size From 1976 to 1978?

Introduction

Bacteria have developed resistance to antibiotics over time. A pesticide that was once highly effective at killing mosquitoes no longer works. House sparrows that live in the northern United States and Canada are larger-bodied than the ones that live in the southern United States and Mexico. These cases are all examples of *microevolution*, or evolutionary change on a small scale. Microevolution occurs within a population. A population is a group of organisms that live in the same area and mate with each other. Biologists define microevolution as a change in the frequency of one or more genes within a population over time. As specific genes within a population become more or less common over time, the traits that are associated with those genes will also change. There are four basic mechanisms that drive microevolution.

The first mechanism of microevolution is a genetic *mutation*. A mutation in a gene can result in an individual having a new version of a trait. The individual with the new gene can then have offspring with the same gene. The new gene could then become more common in a population over time. However, since mutations are rare and only happen in individuals, this process alone cannot result in a big change in the frequency of a gene within a population in only one or two generations.

The second mechanism of microevolution is the process of *migration*. Individuals can either join a population (immigration) or leave a population (emigration). A specific version of a gene will become less common within a population when several individuals with that gene leave the population, and a specific version of a gene will become more common within a population when several individuals with that gene join the population. The migration of a large number of individuals into or out of a population can therefore result in a dramatic shift in the frequency of a gene within a population in a relatively short period of time.

The third mechanism of microevolution is *natural selection*, which occurs when (a) there is variation in a trait among the individuals that make up a population, (b) the trait is determined by one or more genes, (c) the trait affects survival and/or ability to reproduce, and (d) individuals who reproduce pass on their genes to the next generation. The frequency of a gene in any given generation, as a result, reflects the traits and genes of the individuals that were able to survive long enough to reproduce in the previous generation. Over time, genes that determine traits that are associated with an increased chance of survival and successful reproduction will become more common in a population, and genes that determine traits that decrease an individual's chance of survival or reproduction will become less common.

FIGURE L18.1

The Galápagos archipelago

FIGURE L18.2

Daphne Major

The fourth, and final, mechanism of microevolution is *genetic drift*. In any generation, some individuals may, just by chance, survive longer or leave behind more offspring than other individuals. The frequency of a gene in the next generation will therefore reflect the genes and traits of these lucky individuals rather than individuals with traits that are advantageous in terms of survival or reproduction.

It is often difficult to determine which of these four mechanisms is responsible for an evolutionary change within a population. To illustrate this point, you will be studying a population of birds called the medium ground finch (*Geospiza fortis*) that lives in the Galápagos Islands, an archipelago made up of a small group of islands located 600 miles off the coast of mainland Ecuador in South America (see Figure L18.1). There is a small island in the Galápagos called Daphne Major (see Figure L18.2).

Biologists Peter and Rosemary Grant have been studying the medium ground finch population on Daphne Major since 1974. They travel to Daphne Major every summer to study these birds. They capture, tag, and measure the physical characteristics of every bird on the island. They also keep track of the ones that die. Finally, and most importantly, they keep track of when a bird breeds, how many offspring it produces, and how many of those offspring survive long enough to breed.

In the summer of 1976, there were 751 finches on Daphne Major when the Grants left the island. The 1976 medium ground finch population had an average beak depth of 9.65 mm and an average beak length of 10.71 mm. In 1977 a severe drought began, and only 20 mm of rain fell on the island over the entire year. Much of the plant life on the island withered and died. The medium ground finches on Daphne Major, as a result, struggled to find food, and the population quickly decreased in size. By the end of 1978, there were only 90 finches left on the island. When the Grants returned to Daphne Major in 1978 to study the

Environmental Change and Evolution

Which Mechanism of Microevolution Caused the Beak of the Medium Ground Finch Population on Daphne Major to Increase in Size From 1976 to 1978?

characteristics of the finch population, they made an unexpected discovery. They found that the average size of the beak for the medium ground finch on this island had increased. The 1978 population of the medium ground finch population on Daphne Major had an average beak depth of 10.55 mm and an average beak length of 11.61 mm, which was almost a full mm thicker and longer than the 1976 population. The beak of the medium ground finch population had clearly evolved in only two years.

The dramatic increase in the size of the medium ground finch beak was a clear example of microevolution. The Grants therefore wanted to determine which mechanism of microevolution caused the dramatic change in beak size. After they had analyzed the data that they had collected from 1976 to 1978, the Grants proposed that natural selection was the mechanism that caused the beak of the medium ground finch to increase in size. Some scientists, however, thought that this explanation was unacceptable because the change in the trait happened in only two years, and they viewed natural selection as a slow and gradual process. These scientists suggested that a better explanation for the increase in beak size was migration or genetic drift. In this investigation, you will use the data that the Grants collected on Daphne Major to determine which of these three explanations is the most valid or acceptable.

Your Task

Use the Grant's finch data set and what you know about migration, natural selection, and genetic drift to determine which of these three mechanisms of microevolution caused the average size of the medium ground finch beak to increase from 1976 to 1978.

The guiding question of this investigation is, **Which mechanism of microevolution caused the beak of the medium ground finch population on Daphne Major to increase in size from 1976 to 1978?**

Materials

You will use an Excel file called Finch Data during this investigation.

Safety Precautions

Follow all normal lab safety rules.

Investigation Proposal Required?   ☐ Yes   ☐ No

Getting Started

You will need to examine the characteristics of the medium ground finch on Daphne Major before and after the drought of 1977 in order to answer the guiding question for this investigation. Luckily, we know a lot about the physical characteristics of all the medium ground finches found on Daphne Major.

LAB 18

FIGURE L18.3

A medium ground finch

The medium ground finch is a small brown bird (see Figure L18.3). Their brown color helps them blend into their surroundings and avoid the owls that live on the island. (Owls eat small birds.) As with any species, no two medium ground finches are exactly alike. Medium ground finches weigh between 12 and 17 grams and have wings that range in size from 60 mm to 70 mm. These birds also have small beaks. The beak of a medium ground finch ranges in size from 8 mm to 13 mm. The medium ground finch eats seeds (which they must crack open before eating) and the occasional insect.

You may also need to examine the characteristics of the plant life found on Daphne Major before, during, and after the drought of 1977. There are two species of seed-producing plants on Daphne Major: *Tribulus terrestris* (puncturevine) and *Portulaca oleracea* (purslane). The *Tribulus* plants produce large, hard seeds (Figure L18.4) and the *Portulaca* plants produce small, soft seeds (Figure L18.5). Medium ground finches tend to eat seeds from the *Portulaca* plants because they are soft and easy to get.

FIGURE L18.4

Seeds produced by the Tribulus plant

FIGURE L18.5

Seeds produced by the Portulaca plant

You will be given the observations and measurements collected by the Grants. These data have been entered into an Excel spreadsheet. The spreadsheet will make it easier for you to analyze all the available data. To answer the guiding question for this investigation, however, you must determine what type of data you will need to examine and how you will analyze it. To determine *what data you will need to examine and how you will analyze these data*, think about the following questions:

National Science Teachers Association

- What would you expect to see if the change in beak size in the 1976 and 1978 populations of the medium ground finch was caused by migration? Natural selection? Genetic drift?

- What types of comparisons will you need to make between the two populations to test each of the three explanations?

- Are there trends or relationships that you will need to look for in the data?

- Are there other factors that may help you test each explanation?

Connections to Crosscutting Concepts, the Nature of Science, and the Nature of Scientific Inquiry

As you work through your investigation, be sure to think about

- the important role that conceptual models play in science,

- the relationship between structure and function in nature,

- the different types of methods that scientists use to answer questions, and

- the difference between laws and theories in science.

Initial Argument

Once your group has finished collecting and analyzing your data, you will need to develop an initial argument. Your argument must include a claim, evidence to support your claim, and a justification of the evidence. The claim is your group's answer to the guiding question. The evidence is an analysis and interpretation of your data. Finally, the justification of the evidence is why your group thinks the evidence matters. The justification of the evidence is important because scientists can use different kinds of evidence to support their claims. Your group will create your initial argument on a whiteboard. Your whiteboard should include all the information shown in Figure L18.6.

FIGURE L18.6

Argument presentation on a whiteboard

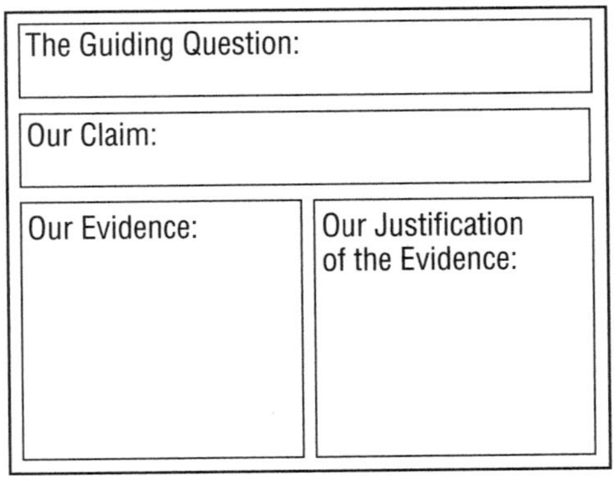

The Guiding Question:	
Our Claim:	
Our Evidence:	Our Justification of the Evidence:

Argumentation Session

The argumentation session allows all of the groups to share their arguments. One member of each group will stay at the lab station to share that group's argument, while the other members of the group go to the other lab stations one at a time to listen to and critique the arguments developed by their classmates. This is similar to how scientists present their arguments to other scientists at conferences. If you are responsible for critiquing your classmates' arguments, your goal is to look for mistakes so these mistakes can be fixed and they can make their argument better. The argumentation session is also a good time to think about ways you can

LAB 18

make your initial argument better. Scientists must share and critique arguments like this to develop new ideas.

To critique an argument, you might need more information than what is included on the whiteboard. You will therefore need to ask the presenter lots of questions. Here are some good questions to ask:

- What did your group do to analyze the data? Why did your group decide to analyze it that way?
- What other ways of analyzing and interpreting the data did your group talk about?
- Why did your group decide to present your evidence in that way?
- Why did your group abandon the other explanations?
- How sure are you that your group's claim is accurate? What could you do to be more certain?

Once the argumentation session is complete, you will have a chance to meet with your group and revise your initial argument. Your group might need to gather more data as part of this process. Remember, your goal at this stage of the investigation is to develop the best argument possible.

Report

Once you have completed your research, you will need to prepare an investigation report that consists of three sections that provide answers to the following questions:

1. What question were you trying to answer and why?
2. What did you do during your investigation and why did you conduct your investigation in this way?
3. What is your argument?

Your report should answer these questions in two pages or less. The report must be typed, and any diagrams, figures, or tables should be embedded into the document. Be sure to write in a persuasive style; you are trying to convince others that your claim is acceptable or valid!

Checkout Questions

Lab 18. Environmental Change and Evolution: Which Mechanism of Microevolution Caused the Beak of the Medium Ground Finch Population on Daphne Major to Increase in Size From 1976 to 1978?

Use the following information to answer questions 1–3.

The beach mouse (*Peromyscus polionotus*), shown in the figure below, is a small rodent found in the southeastern United States. It lives primarily in old fields and on white sand beaches. The fur of the beach mouse ranges from dark brown to very light brown. The darkest-color mice tend to live inland, and the lighter-color mice tend to live on light sand beaches.

A dark brown beach mouse

Some scientists think the trend in the coloration of the beach mouse is due to natural selection, and others think it is due to genetic drift.

1. Describe the process of natural selection, and explain how this process could result in darker-color mice living inland and lighter-color mice living on light sand beaches.

LAB 18

2. Describe the process of genetic drift and explain how this process could result in darker-color mice living inland and lighter-color mice living on light sand beaches.

3. Describe a test that you could conduct to determine if pattern in mouse coloration is due to natural selection or genetic drift.

4. Scientists often use existing models or develop a new model to help understand a system. Explain why models are useful in science, using an example from your investigation about environmental change and evolution.

National Science Teachers Association

5. The structures that make up an organism's body are not related to the functions they perform.

   a. I agree with this statement.
   b. I disagree with this statement.

   Explain your answer, using an example from your investigation about environmental change and evolution.

6. A scientific law describes the behavior of a natural phenomenon, and a scientific theory is a well-substantiated explanation of some aspect of the natural world.

   a. I agree with this statement.
   b. I disagree with this statement.

   Explain your answer, using an example from your investigation about environmental change and evolution.

7. There is no universal step-by-step scientific method that all scientists follow; rather, the choice of method depends on the objectives of the research. Explain why scientists need to use different types of methods to answer different types of questions, using an example from your investigation about environmental change and evolution.

LAB 19

Teacher Notes

Lab 19. Phylogenetic Trees and the Classification of Fossils: How Should Biologists Classify the Seymouria?

Purpose

The purpose of this lab is for students to *apply* what they have learned about phylogenetic trees and phylogenetic classification to classify an extinct organism. This lab gives students an opportunity to work with actual skeletons and replica fossils. Through this activity, students will learn about the importance of looking for patterns in nature and about the relationship between structure and function in organisms. Students will also have the opportunity to reflect on the difference between observations and inferences in science and how science as a body of knowledge develops over time.

The Content

Biologists use *phylogenetic trees* to represent evolutionary relationships between species. A phylogenetic tree is a branching diagram that shows how various species are related to each other based on similarities and differences in their physical and/or genetic characteristics. The root of a phylogenetic tree represents a common ancestor, and the tips of the branches represent the descendants (Figure 19.1). When biologists develop these trees, the root is often a point of much debate, particularly since several lines of evidence, especially fossils, may not have identified a specific common ancestor. The use of outlier groups, which are species that are not closely related to the others in an analysis, help to create a pattern that has a root when no specific common ancestor is known. By comparing the rest of the species with the outlier group, a biologist creates the first branching point, which helps to "root" the tree. That root point can also help the biologist connect the tree he or she develops to other trees developed by other biologists.

As you move from the tips to the root of the tree, you are moving backward in time. The forks in the tree represent *speciation events*. When a speciation event occurs, a single species (or an *ancestral lineage*) gives rise to two new species (or two *daughter lineages*). These speciation events can include genetic and environmental events that lead to the reproductive isolation of different groups of related species. These events can include the development of major genetic differences (through mutation) that stop two groups from continuing to reproduce with each other. One group of a species may also become geographically isolated from others and then change over time in a different way than the populations from which they were separated. Each species in a phylogenetic tree has a part of its history that is unique to that species and parts that are shared with other species (Figure 19.2). Similarly,

FIGURE 19.1

Components of a phylogenetic tree

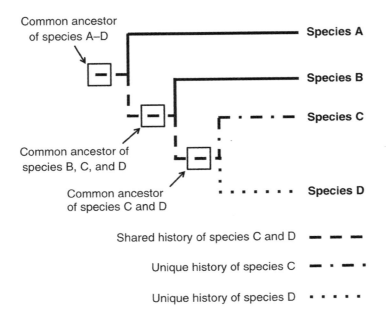

FIGURE 19.2

Evolutionary history in a phylogenetic tree

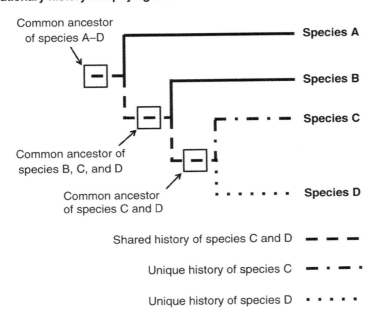

each species has ancestors that are unique to that species and ancestors that are shared with other species (see species C and D compared with species A and B in Figure 19.2, p. 315).

A *clade* is a grouping of species that includes a common ancestor and all the descendants (living and extinct) of that ancestor. Clades are nested within one another—biologists call this a *nested hierarchy.* A clade may include thousands of species or just a few. Some examples of clades at different levels are marked in the phylogenetic tree shown in Figure 19.3. Notice how clades are nested within larger clades.

FIGURE 19.3

Clades with a phylogenetic tree

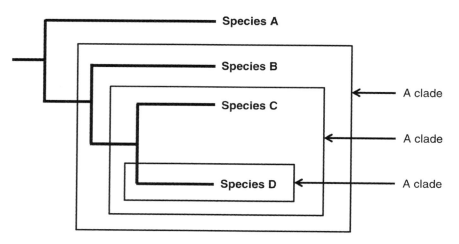

To create a phylogenetic tree, biologists collect data about the characters of organisms. Characters are heritable traits, such as physical features or genetic sequences, which can be compared across organisms. Biologists begin by examining representatives of each lineage to learn about their physical features, and they tend to look for specific features called *shared derived characters* when they create a phylogenetic tree. A shared derived character is one that developed or appeared at some point in the evolutionary history of an organism and is shared by other closely related organisms but not by distantly related organisms. Biologists then use these shared derived characters to group the organisms into less and less inclusive clades. Similarly, genetic sequences among common genes present in many living things will be compared with one another across species. The differences between these sequences are taken into account and biologists use special computer models to determine the relationships between the species. The computer models take into account the probability of certain differences and how much some types of mutations and genetic groupings occur.

In addition to using phylogenetic trees to represent evolutionary relationships between organisms, biologists also use them as a system of classification. The phylogenetic classification system classifies species by clades rather than assigning every one to a kingdom, phylum, class, order, family, and genus like the Linnaean system of classification. Figure 19.4 provides the evolutionary history of some major types of vertebrates and includes the names of various clades that are used to classify vertebrate species. The Linnaean classification system is based primarily on evidence from physical traits and behaviors. Clade analysis relies more significantly on genetic and protein structure information, but still also accounts for fossil and behavioral data. The two different classification systems often agree with each other. However, when they do not, the system that is used depends on the purpose of the biologists and who will use the outcomes of their analysis. Some investigations intend to focus on the classification focus of clade diagrams. Other investigations use the clade structures to support arguments related to comparisons across many species and their relation to other evolutionary events and phenomena.

FIGURE 19.4

Vertebrate clades

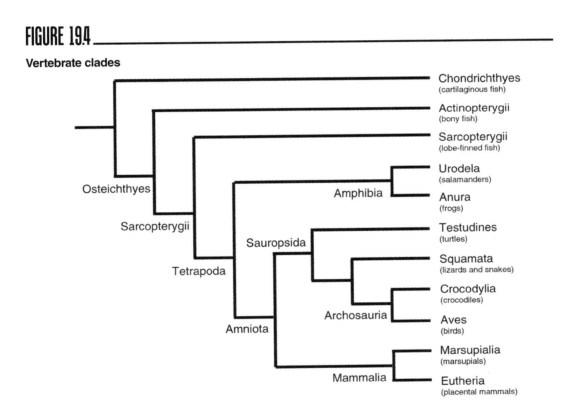

When biologists find a fossil of an extinct organism, they can use phylogenetic trees to understand how it is related to other organisms and to help classify it. The classification of an extinct organism is often difficult, however, because fossils only provide limited information about the characteristics of the organism. Thus, other lines of evidence, including any genetic information that can be found related to the organism, are especially helpful in creating better understandings of organisms' evolutionary history. As biologists continue to investigate the evolutionary relationships among organisms, they will also discover and develop further sources of evidence. As this new evidence emerges, the phylogenetic trees will increase in their complex structure, connectedness to other trees, and reliability.

Timeline

The instructional time needed to implement this lab investigation is 180–250 minutes. Appendix 2 (p. 355) provides options for implementing this lab investigation over several class periods. Option E (250 minutes) should be used if students are unfamiliar with scientific writing, because this option provides extra instructional time for scaffolding the writing process. You can scaffold the writing process by modeling, providing examples, and providing hints as students as write each section of the report. Option F (180 minutes) should be used if students are familiar with scientific writing and have the skills needed to write an investigation report on their own. In option F, students complete stage 6 (writing the investigation report) and stage 8 (revising the investigation report) as homework.

Materials and Preparation

The materials needed to implement this investigation are listed in Table 19.1. A replica of a Seymouria fossil can be purchased from The Dinosaur Store (*www.dinosaurstore.com*). The frog, lizard, pigeon, bat, and rat skeletons can be purchased from Skulls Unlimited (*www.skullsunlimited.com*). Other science supply companies, such as Carolina, Flinn Scientific, and Ward's Science, might sell skeletons and a replica Seymouria fossil at lower prices. One replica fossil and one skeleton for each animal is all that is needed to implement this lab because each group can be given one specimen at a time. Once the group is done documenting the various features of one specimen, they can pass it on to the next group.

The University College London's Vertebrate Paleontology and Evolution program also has numerous images of vertebrates available at *www.ucl.ac.uk/museums-static/obl4he/ vertebratediversity/index.html*. However, it is difficult to see and document many of the different features of skeletons and fossils when images are used, so we recommend using actual specimens for this lab.

TABLE 19.1

Materials list

Item	Quantity
Seymouria fossil	At least 1 per class
Frog skeleton	At least 1 per class
Lizard skeleton	At least 1 per class
Pigeon skeleton	At least 1 per class
Bat skeleton	At least 1 per class

Table 19.1 *(continued)*

Item	Quantity
Rat skeleton	At least 1 per class
Investigation Proposal C (optional)	1 per group
Whiteboard, 2' × 3'*	1 per group
Lab Handout	1 per student
Peer-review guide	1 per student
Checkout Questions	1 per student

* Students can also use computer and presentation software such as Microsoft PowerPoint or Apple Keynote to create their arguments.

Safety Precautions

Follow all normal lab safety rules. In addition, take the following safety precautions:

1. Use caution in handling skeletons. They can have sharp edges, which can cut skin.

2. Wash hands with soap and water after completing the lab activity.

Topics for the Explicit and Reflective Discussion

Concepts That Can Be Used to Justify the Evidence

To provide an adequate justification of their evidence, students must explain why they included the evidence in their arguments and make the assumptions underlying their analysis and interpretation of the data explicit. In this investigation, students can use the following concepts to help justify their evidence:

- Phylogenetic trees and the assumptions underlying their construction
- Phylogenetic classification
- Shared derived characters

We recommend that you review these concepts during the explicit and reflective discussion to help students make this connection.

How to Design Better Investigations

It is important for students to reflect on the strengths and weaknesses of the investigation they designed during the explicit and reflective discussion. Students should therefore be encouraged to discuss ways to eliminate potential flaws, measurement errors, or sources of bias in their investigations. To help students be more reflective about the design of their investigation, you can ask the following questions:

- What were some of the strengths of your investigation? What made it scientific?
- What were some of the weaknesses of your investigation? What made it less scientific?
- If you were to do this investigation again, what would you do to address the weaknesses in your investigation? What could you do to make it more scientific?

Crosscutting Concepts

This investigation is aligned with two crosscutting concepts found in *A Framework for K–12 Science Education,* and you should review these concepts during the explicit and reflective discussion.

- *Patterns:* Scientists look for patterns in nature and attempt to understand the underlying cause of these patterns. Biologists, for example, look for patterns in the ways embryos grow and develop and then compare these patterns across different types of animals.
- *Structure and function:* In nature, the way a living thing is shaped or structured determines how it functions and places limits on what it can and cannot do.

The Nature of Science and the Nature of Scientific Inquiry

This investigation is aligned with two important concepts related to the *nature of science* (NOS) and the *nature of scientific inquiry* (NOSI), and you should review these concepts during the explicit and reflective discussion.

- *The difference between observations and inferences:* An observation is a descriptive statement about a natural phenomenon, whereas an inference is an interpretation of an observation. Students should also understand that current scientific knowledge and the perspectives of individual scientists guide both observations and inferences. Thus, different scientists can have different but equally valid interpretations of the same observations due to differences in their perspectives and background knowledge.
- *Changes in scientific knowledge over time:* A person can have confidence in the validity of scientific knowledge but must also accept that scientific knowledge may be abandoned or modified in light of new evidence or because existing evidence has been reconceptualized by scientists. There are many examples in the history of science of both evolutionary changes (i.e., the slow or gradual refinement of ideas) and revolutionary changes (i.e., the rapid abandonment of a well-established idea) in scientific knowledge.

Hints for Implementing the Lab

- Be sure to allow each group to look at a specimen as part of the tool talk before they design their investigation. Students will be able to design a better investigation if they understand what information they can and cannot get from the specimens.

- If students are struggling with types of observations to make, you can provide them with the information in Table 19.2, which lists some shared derived characters of each clade.

- Encourage students to think of ways to limit measurement error before they begin collecting their data.

- This lab can be made more challenging by requiring students to examine a larger sample of skeletons.

TABLE 19.2

Shared derived characters of four clades

Anura	Squamata
• Elongated hind limbs, including the anklebones (tarsals) and foot bones (metatarsals and phalanges) • Short, stiff vertebral column (nine or fewer vertebrae proper) and no ribs • Short and flat head • No teeth on the dentary • Fused radius and ulna to form a compound radio-ulna	• Cranial kinesis—a high degree of flexibility between the bones of the back of the skull, allowing relative movements between them • Pleurodont dentition—teeth set into the side of the inner surfaces of the jaws • Loss of gastralia (ventral belly ribs) • Double-hooked fifth metatarsal, functionally analogous to the mammalian heel
Aves	**Eutherians (Placental)**
• Wings, formed of the humerus, radius, ulna, wrist, and three digits • Fused clavicles, forming the furcula (wishbone) • Large keeled sternum (breastbone) • No teeth; replaced by a horny beak • Caudal vertebrae reduced and fused to form the pygostyle, which supports the tail feathers • Reversed hallux (big toe) on feet, specialized for perching	• The presence of a malleolus at the bottom of the fibula (the smaller of the two shin bones) • The complete mortise and tenon upper ankle joint, where the rearmost bones of the foot fit into a socket formed by the ends of the tibia and fibula • A wide opening at the bottom of the pelvis, which allows the birth of large, well-developed offspring

Topic Connections

Table 19.3 (p. 322) provides an overview of the scientific practices, crosscutting concepts, disciplinary core ideas, and support ideas at the heart of this lab investigation. In addition, it lists NOS and NOSI concepts for the explicit and reflective discussion. Finally, it lists

LAB 19

literacy and mathematics skills (*CCSS ELA* and *CCSS Mathematics*) that are addressed during the investigation.

TABLE 19.3

Lab 19 alignment with standards

Scientific practices	• Asking questions and defining problems • Planning and carrying out investigations • Analyzing and interpreting data • Constructing explanations • Engaging in argument from evidence • Obtaining, evaluating, and communicating information
Crosscutting concepts	• Patterns • Structure and function
Core ideas	• LS4: Biological evolution: Unity and diversity
Supporting ideas	• Descent with modification • Phylogenetic trees and the assumptions underlying their construction • Phylogenetic classification • Shared derived characters
NOS and NOSI concepts	• Observations and inferences • Changes in scientific knowledge over time
Literacy connections (*CCSS ELA*)	• *Reading:* Key ideas and details, craft and structure, integration of knowledge and ideas • *Writing:* Text types and purposes, production and distribution of writing, research to build and present knowledge, range of writing • *Speaking and listening:* Comprehension and collaboration, presentation of knowledge and ideas
Mathematics connections (*CCSS Mathematics*)	• Make sense of problems and persevere in solving them • Reason abstractly and quantitatively • Construct viable arguments and critique the reasoning of others • Model with mathematics • Use appropriate tools strategically • Look for and express regularity in repeated reasoning

Notes

The phylogenetic tree of vertebrates and the classification of vertebrates used in this lab are based on information found at University College London's Vertebrate Paleontology and Evolution program: *www.ucl.ac.uk/museums-static/obl4he/vertebratediversity/index.html.*

Lab Handout

Lab 19. Phylogenetic Trees and the Classification of Fossils: How Should Biologists Classify the Seymouria?

Introduction

Biologists use *phylogenetic trees* to represent evolutionary relationships between species. A phylogenetic tree is a branching diagram that shows how various species are related to each other based on similarities and differences in their physical and/or genetic characteristics. The root of a phylogenetic tree represents a common ancestor, and the tips of the branches represent the descendants (Figure L19.1).

FIGURE L19.1

Components of a phylogenetic tree

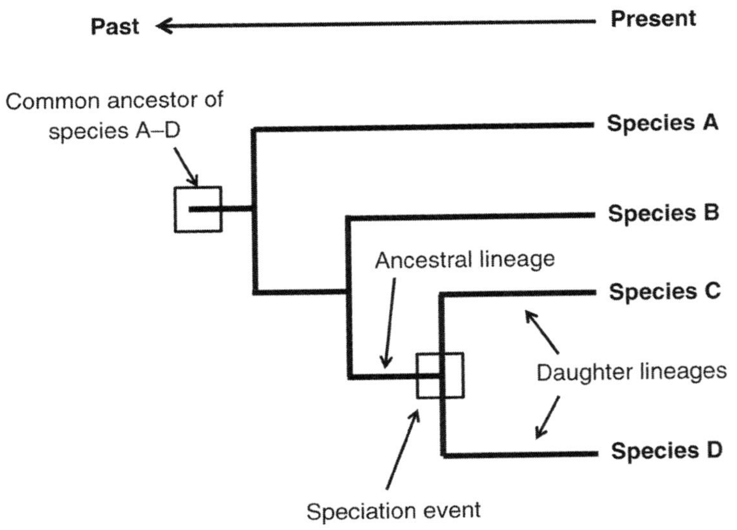

As you move from the tips to the root of the tree, you are moving backward in time. The forks in the tree represent *speciation events*. When a speciation event occurs, a single species (or an *ancestral lineage*) gives rise to two new species (or two *daughter lineages*). Each species in a phylogenetic tree has a part of its history that is unique to that species and parts that are shared with other species (Figure L19.2, p. 324). Similarly, each species has ancestors that are unique to that species and ancestors that are shared with other species (see species C and D compared with species A and B in Figure L19.2).

FIGURE L19.2

Evolutionary history in a phylogenetic tree

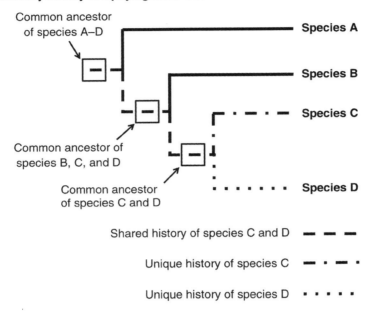

A *clade* is a grouping of species that includes a common ancestor and all the descendants (living and extinct) of that ancestor. Clades are nested within one another—biologists call this a *nested hierarchy*. A clade may include thousands of species or just a few. Some examples of clades at different levels are marked in the phylogenetic tree shown in Figure L19.3. Notice how clades are nested within larger clades.

FIGURE L19.3

Clades with a phylogenetic tree

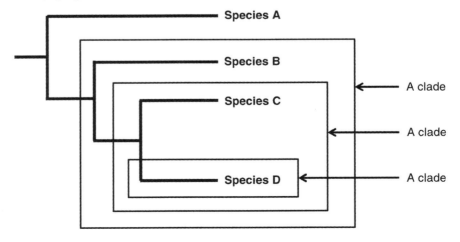

To create a phylogenetic tree, biologists collect data about the characters of organisms. Characters are heritable traits, such as physical features or genetic sequences, which can be compared across organisms. Biologists begin by examining representatives of each lineage to learn about their physical features, and they tend to look for specific features called *shared derived characters* when they create a phylogenetic tree. A shared derived character is one that developed or appeared at some point in the evolutionary history of an organism and is shared by other closely related organisms but not by distantly related organisms. Biologists then use these features to group the organisms into less and less inclusive clades.

In addition to using phylogenetic trees to represent evolutionary relationships between organisms, biologists also use them as a system of classification. The phylogenetic classification system classifies species by clades rather than assigning every one to a kingdom, phylum, class, order, family, and genus like the Linnaean system of classification. Figure L19.4 provides the evolutionary history of some major types of vertebrates and includes the names of various clades that are used to classify vertebrate species.

FIGURE L19.4

Vertebrate clades

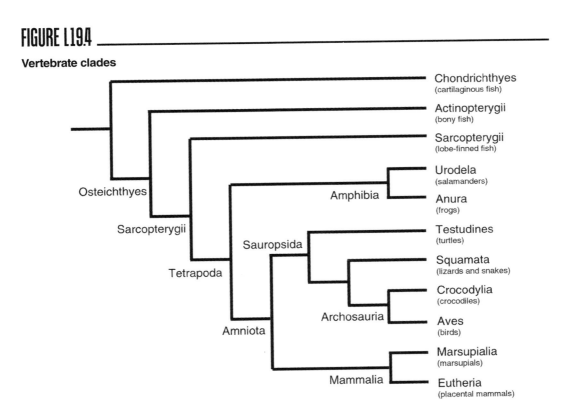

When biologists find a fossil of an extinct organism, they can use phylogenetic trees to understand how it is related to other organisms and to help classify it. The classification of an extinct organism is often difficult, however, because fossils only provide limited

LAB 19

Seymouria

information about the characteristics of the organism. To illustrate this point, you will attempt to classify an organism called the Seymouria (*Seymouria baylorensis*; Figure L19.5). The first fossil of the Seymouria was found in the early 1900s in Baylor County, Texas. The Seymouria was about 60 cm long and lived during the Permian Period (between 299 and 251 million years ago) throughout North America and Europe. The Seymouria has many features that make it difficult to classify. In this investigation, you will have a chance to examine the features of the Seymouria and the features of some representative species from different clades and then attempt to classify it.

Your Task

Use skeletons from vertebrates belonging to four clades (Anura, Squamata, Aves, and Eutheria), what you know about phylogenetic classification, and the characteristics of vertebrate clades to classify the Seymouria.

The guiding question of this investigation is, **How should biologists classify the Seymouria?**

Materials

You may use any of the following materials during your investigation:

- Seymouria fossil
- Frog skeleton (Anura)
- Lizard skeleton (Squamata)
- Pigeon skeleton (Aves)
- Bat skeleton (Eutheria)
- Rat skeleton (Eutheria)

Safety Precautions

Follow all normal lab safety rules. In addition, take the following safety precautions:

1. Use caution in handling skeletons. They can have sharp edges, which can cut skin.

2. Wash hands with soap and water after completing the lab activity.

Investigation Proposal Required?   ☐ Yes   ☐ No

Getting Started

To answer the guiding question, you will need to compare and contrast the features of the five modern vertebrate skeletons with the Seymouria fossil. You will be supplied with either actual specimens or images of them. You must determine what type of data you need to collect from these specimens, how you will collect the data, and how you will analyze the data. To determine *what type of data you need to collect*, think about the following questions:

- Which of the modern animals are more closely related to each other?
- What are the characteristics that biologists use to group organisms into vertebrate clades?
- Which characteristics of the specimens will you need to examine?
- How many different characteristics of the specimens will you need to examine?

To determine *how you will collect the data*, think about the following questions:

- How will you quantify differences and similarities in specimens?
- How will you make sure that your data are of high quality?
- What will you do with the data you collect?

To determine *how you will analyze your data*, think about the following questions:

- How will you compare and contrast the various specimens?
- What type of graph or table could you create to help make sense of your data?

Connections to Crosscutting Concepts, the Nature of Science, and the Nature of Scientific Inquiry

As you work through your investigation, be sure to think about

- the importance of looking for patterns in science,
- the relationship between structure and function in nature,
- the difference between observations and inferences in science, and
- how scientific knowledge changes over time.

Initial Argument

Once your group has finished collecting and analyzing your data, you will need to develop an initial argument. Your argument must include a claim, evidence to support your claim, and a justification of the evidence. The claim is your group's answer to the guiding question. The evidence is an analysis and interpretation of your data. Finally, the justification of the evidence is why your group thinks the evidence matters. The justification of the evidence is important because scientists can use different kinds of

LAB 19

Argument presentation on a whiteboard

The Guiding Question:	
Our Claim:	
Our Evidence:	Our Justification of the Evidence:

evidence to support their claims. Your group will create your initial argument on a whiteboard. Your whiteboard should include all the information shown in Figure L19.6.

Argumentation Session

The argumentation session allows all of the groups to share their arguments. One member of each group will stay at the lab station to share that group's argument, while the other members of the group go to the other lab stations one at a time to listen to and critique the arguments developed by their classmates. This is similar to how scientists present their arguments to other scientists at conferences. If you are responsible for critiquing your classmates' arguments, your goal is to look for mistakes so these mistakes can be fixed and they can make their argument better. The argumentation session is also a good time to think about ways you can make your initial argument better. Scientists must share and critique arguments like this to develop new ideas.

To critique an argument, you might need more information than what is included on the whiteboard. You will therefore need to ask the presenter lots of questions. Here are some good questions to ask:

- What did your group do to analyze the data? Why did your group decide to analyze it that way?
- What other ways of analyzing and interpreting the data did your group talk about?
- Why did your group decide to present your evidence in that way?
- Why did your group abandon the other explanations?
- How sure are you that your group's claim is accurate? What could you do to be more certain?

Once the argumentation session is complete, you will have a chance to meet with your group and revise your initial argument. Your group might need to gather more data as part of this process. Remember, your goal at this stage of the investigation is to develop the best argument possible.

Report

Once you have completed your research, you will need to prepare an investigation report that consists of three sections that provide answers to the following questions:

1. What question were you trying to answer and why?

2. What did you do during your investigation and why did you conduct your investigation in this way?

3. What is your argument?

Your report should answer these questions in two pages or less. The report must be typed, and any diagrams, figures, or tables should be embedded into the document. Be sure to write in a persuasive style; you are trying to convince others that your claim is acceptable or valid!

Checkout Questions

Lab 19. Phylogenetic Trees and the Classification of Fossils: How Should Biologists Classify the Seymouria?

1. What is a phylogenetic tree? What is a clade?

2. This diagram shows the evolutionary relationships among several major groups of organisms. Using your knowledge of clades, identify the pairs of organisms that are *most* closely related and the pair that is *least* closely related. How many clades are there in this diagram?

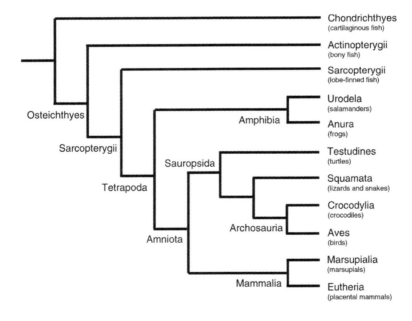

3. In science, observations and inferences are the same thing.

   a. I agree with this statement.
   b. I disagree with this statement.

   Explain your answer, using an example from your investigation about phylogenetic trees and the classification of fossils.

4. Scientific knowledge changes and develops over time.

   a. I agree with this statement.
   b. I disagree with this statement.

   Explain your answer, using an example from your investigation about phylogenetic trees and the classification of fossils.

5. Scientists often look for patterns during their investigations. Explain why patterns are important to look for, using an example from your investigation about phylogenetic trees and the classification of fossils.

6. The relationship between the structure and function of organisms' features is an important area of study in science. Discuss why it is important for scientists to understand this relationship, using an example from your investigation about phylogenetic trees and the classification of fossils.

LAB 20

Lab 20. Descent With Modification and Embryonic Development: Does Animal Embryonic Development Support or Refute the Theory of Descent With Modification?

Purpose

The purpose of this lab is for students to *apply* what they have learned about the theory of descent with modification to explain similarities in animal embryo development. This lab gives students an opportunity to use images of animal embryos to test a fundamental idea in biology. Through this activity, students will learn about the importance of looking for patterns in nature and about the relationship between structure and function in organisms. Students will also have the opportunity to reflect on the way that science as a body of knowledge develops over time and the various methods that scientists can use during an investigation.

The Content

Biological evolution is defined as descent with modification. This definition includes both *microevolution,* which refers to a change in gene frequency in a population from one generation to the next, and *macroevolution,* which refers to the origin of new species. The central idea of descent with modification is that all life on Earth shares a common ancestor. All organisms on Earth, as a result, share some features. Organisms also have *shared derived characters.* A shared derived character is one that developed or appeared at some point in the evolutionary history of an organism and is shared by other closely related organisms but not by distantly related organisms. Shared derived characteristics, such as four limbs or the ability to produce milk, allow biologists to determine how closely related an organism is to other organisms. Two organisms that have many shared derived characters in common are considered to be more closely related than two organisms that have few shared derived characters in common.

Many organisms also go through similar developmental stages. The embryos of many different kinds of animals (mammals, birds, reptiles, fish, etc.) look very similar, and it is often difficult to tell them apart. Many traits of one type of animal also appear in the embryo of another type of animal. For example, fish embryos and chicken embryos both have gill slits. In fish they develop into gills, but in chickens the gill slits disappear before birth. These animals go through similar developmental stages because they share a common ancestor that went through these same stages. Although these animals diverged from this common ancestor over time and gradually evolved different traits, they still share the same basic development process.

Timeline

The instructional time needed to implement this lab investigation is 180–250 minutes. Appendix 2 (p. 355) provides options for implementing this lab investigation over several class periods. Option E (250 minutes) should be used if students are unfamiliar with scientific writing, because this option provides extra instructional time for scaffolding the writing process. You can scaffold the writing process by modeling, providing examples, and providing hints as students as write each section of the report. Option F (180 minutes) should be used if students are familiar with scientific writing and have the skills needed to write an investigation report on their own. In option F, students complete stage 6 (writing the investigation report) and stage 8 (revising the investigation report) as homework.

Materials and Preparation

The materials needed to implement this investigation are listed in Table 20.1. The *Lab 20 Embryo Stages.pptx* file is available at *www.nsta.org/publications/press/extras/adi-lifescience.aspx*. We recommend that you download the file, print out the 80 individual slides as a handout (4 slides per page), cut out each slide, and laminate each one. You can then group the embryo pictures by animal as a set of cards. There are 10 embryo pictures for each of the eight animals, so each set of cards will include 10 cards.

The images of animal embryos at various stages of development found in the Animal Embryos PowerPoint file have been recreated by Jennifer Schellinger from the following sources: Blom and Lilja 2005, Boback, Dichter, and Mistry (2012, Greenbaum 2002; Kimmel et al. 1995; Richardson et al. 2014; Segardell et al. 2008; Tokita 2006.

TABLE 20.1

Materials list

Item	Quantity
Computer with a Microsoft PowerPoint Presentation application	1 per group
Embryo Stages PowerPoint file	At least 1 set per class
Investigation Proposal A (optional)*	1 per group
Whiteboard, 2' × 3'†	1 per group
Lab Handout	1 per student
Peer-review guide	1 per student
Checkout Questions	1 per student

* We highly recommend that students fill out an investigation proposal for this lab.

† Students can also use computer and presentation software such as Microsoft PowerPoint or Apple Keynote to create their arguments.

LAB 20

Safety Precautions

Follow all normal lab safety rules.

Topics for the Explicit and Reflective Discussion

Concepts That Can Be Used to Justify the Evidence

To provide an adequate justification of their evidence, students must explain why they included the evidence in their arguments and make the assumptions underlying their analysis and interpretation of the data explicit. Students will therefore need to highlight specific predictions that are consistent with the theory of descent with modification to justify their evidence. We recommend that you discuss some predictions that are consistent with this theory along with predictions that are associated with creationism (species are fixed entities that neither change nor give rise to different species) and transformism (species become new, more complex species). In addition, students can use the following concepts to help justify their evidence:

- Microevolution
- Macroevolution
- Shared derived characters

We recommend that you review these concepts during the explicit and reflective discussion to help students make this connection.

How to Design Better Investigations

It is important for students to reflect on the strengths and weaknesses of the investigation they designed during the explicit and reflective discussion. Students should therefore be encouraged to discuss ways to eliminate potential flaws, measurement errors, or sources of bias in their investigations. To help students be more reflective about the design of their investigation, you can ask the following questions:

- What were some of the strengths of your investigation? What made it scientific?
- What were some of the weaknesses of your investigation? What made it less scientific?
- If you were to do this investigation again, what would you do to address the weaknesses in your investigation? What could you do to make it more scientific?

Crosscutting Concepts

This investigation is aligned with two crosscutting concepts found in *A Framework for K–12 Science Education,* and you should review these concepts during the explicit and reflective discussion.

- *Patterns:* Scientists look for patterns in nature and attempt to understand the underlying cause of these patterns. Biologists, for example, look for patterns in the ways embryos grow and develop and then compare these patterns across different types of animals.

- *Structure and function:* In nature, the way a living thing is shaped or structured determines how it functions and places limits on what it can and cannot do.

The Nature of Science and the Nature of Scientific Inquiry

This investigation is aligned with two important concepts related to the *nature of science* (NOS) and the *nature of scientific inquiry* (NOSI), and you should review these concepts during the explicit and reflective discussion.

- *Changes in scientific knowledge over time:* A person can have confidence in the validity of scientific knowledge but must also accept that scientific knowledge may be abandoned or modified in light of new evidence or because existing evidence has been reconceptualized by scientists. There are many examples in the history of science of both evolutionary changes (i.e., the slow or gradual refinement of ideas) and revolutionary changes (i.e., the rapid abandonment of a well-established idea) in scientific knowledge.

- *Methods used in scientific investigations:* Examples of methods include experiments, systematic observations of a phenomenon, literature reviews, and analysis of existing data sets; the choice of method depends on the objectives of the research. There is no universal step-by-step scientific method that all scientists follow; rather, different scientific disciplines (e.g., biology vs. physics) and fields within a discipline (e.g., evolutionary biology vs. molecular biology) use different types of methods, use different core theories, and rely on different standards to develop scientific knowledge.

Hints for Implementing the Lab

- You should only need to make one copy of each set of cards because each group of students can focus on one set of cards at a time (assuming you divide the class into eight groups of students). The groups can then switch with another group until they have had a chance to examine all eight organisms.

- Be sure to allow each group to look that the images of the embryos as part of the tool talk before they design their investigation. Students will be able to design a better investigation if they understand what information they can and cannot get from the images.

- If students are struggling with what types of observations to make, here are some suggestions:
  - Presence of four limbs

- Presence of a dorsal (back) nerve cord
- Presence of notochord (a semiflexible rod the runs the length of the animal beneath the dorsal nerve cord)
- Presence of a post-anal tail (or extension of the notochord and nerve cord past the anus)
- Presence of pharyngeal slits (openings between the pharynx or throat and the outside) or arches
- Shape of eye

Topic Connections

Table 20.2 provides an overview of the scientific practices, crosscutting concepts, disciplinary core ideas, and support ideas at the heart of this lab investigation. In addition, it lists NOS and NOSI concepts for the explicit and reflective discussion. Finally, it lists literacy and mathematics skills (*CCSS ELA* and *CCSS Mathematics*) that are addressed during the investigation.

TABLE 20.2

Lab 20 alignment with standards

Scientific practices	• Asking questions and defining problems • Planning and carrying out investigations • Analyzing and interpreting data • Constructing explanations • Engaging in argument from evidence • Obtaining, evaluating, and communicating information
Crosscutting concepts	• Patterns • Structure and function
Core ideas	• LS4: Biological evolution: Unity and diversity
Supporting ideas	• Descent with modification • Microevolution • Macroevolution • Shared derived characters
NOS and NOSI concepts	• Changes in scientific knowledge over time • Methods used in scientific investigations
Literacy connections (*CCSS ELA*)	• *Reading:* Key ideas and details, craft and structure, integration of knowledge and ideas • *Writing:* Text types and purposes, production and distribution of writing, research to build and present knowledge, range of writing • *Speaking and listening:* Comprehension and collaboration, presentation of knowledge and ideas
Mathematics connections (*CCSS Mathematics*)	• Make sense of problems and persevere in solving them • Reason abstractly and quantitatively • Construct viable arguments and critique the reasoning of others • Use appropriate tools strategically

References

Blom, J., and C. Lilja. 2005. A comparative study of embryonic development of some bird species with different patterns of postnatal growth. *Zoology* 108: 81–95.

Boback, S., E. Dichter, and H. Mistry. 2012. A developmental staging series for the African house snake, *Boaedon (Lamprophis) fuliginosus*. *Zoology* 115: 38–46.

Greenbaum, E. 2002. A standardized series of embryonic stages for the emydid turtle *Trachemys scripta*. *Canadian Journal of Zoology* 80: 1350–1370.

Kimmel, C., W. Ballard, S. Kimmel, B. Ullmann, and T. Schilling. 1995. Stages of embryonic development of the zebrafish. *Developmental Dynamics* 203: 253–310.

Richardson L., S. Venkataraman, P. Stevenson, Y. Yang, J. Moss, L. Graham, N. Burton, B. Hill, J. Rao, R. A. Baldock, and C. Armit. 2014. *EMAGE mouse embryo spatial gene expression database: 2014 update. Nucleic Acids Research* 42: D835-D844.

Segerdell, E., J. B. Bowes, N. Pollet, and P. D. Vize. 2008. An ontology for *Xenopus* anatomy and development. *BMC Developmental Biology* 8: 92.

Tokita, M. 2006. Normal embryonic development of the Japanese pipistrelle, *Pipistrellus abramus*. *Zoology* 109: 137–147.

Lab Handout

Lab 20. Descent With Modification and Embryonic Development: Does Animal Embryonic Development Support or Refute the Theory of Descent With Modification?

Introduction

One of Charles Darwin's most revolutionary ideas was that all living things are related. According to Darwin, all organisms found on Earth are related to each other because they all share a common ancestor. He argued that this common ancestor lived on Earth sometime in the distant past but is now extinct. All living things, as a result, share many of the same features, and the differences we see in organisms are simply the result of gradual modifications to these features over long periods of time.

Darwin first came to this conclusion by examining similarities and differences in the traits of closely related animals, such as the beaks of the Galápagos finches in Figure L20.1. To explain the similarities in the beaks of these birds, Darwin suggested that the birds were all the descendants of the same ancestor that originally colonized the Galápagos Islands, and the differences in their beaks were simply the result of gradual modifications in beak shape over many generations. The modifications in beak shape made the birds better adapted to survive in a particular environment. He called this theory *descent with modification*. He argued that natural selection, over time, could slowly select for or against slight variations in the basic shape of the beaks of the birds. This selection process would gradually result in some birds with thick beaks that are able to crack nuts and some birds with narrow beaks that are able to pick insects out of the bark of trees.

The theory of descent with modification can also be used to explain the existence of *homologous structures*, such as the limbs of the four animals in Figure L20.2. Homologous structures are parts of organisms that have similar components even though they may have very different functions. To explain the similar bone structure in these animals, Darwin once again argued that these animals share a common ancestor that had a limb that consisted of a humerus, an ulna, a radius, and carpals and that the observed differences in structure are simply the result of the process of natural selection. Over time, the process of natural selection gradually changed the shape of individual bones but did not completely change the basic layout of the limb. This selection process eventually resulted in whale fins and bird wings that had fingers similar to the fingers of a human or dog. These variations in structure would give their owners an advantage in a particular environment, such as the air in the case of the bird or the ocean in the case of the whale.

FIGURE L20.1

Differences in the beaks of finches found in the Galápagos Islands (Darwin 1845)

1. Geospiza magnirostris.
3. Geospiza parvula.

2. Geospiza fortis.
4. Certhidea olivacea.

FIGURE L20.2

Examples of homologous structures: the limbs of a human, dog, bird, and whale

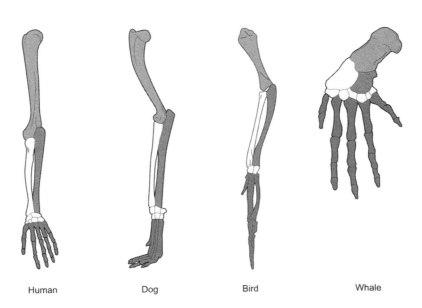

Human   Dog   Bird   Whale

An important principle of descent with modification is that two closely related species will have more features in common than two species that are not as closely related. For example, two species of bird will have more features in common than a bird and an amphibian. Closely related species have more features in common because they shared a common ancestor in the more recent past. More time gives natural selection more opportunities to modify features or to produce new features in each independent lineage.

The theory of descent with modification can explain many different observations, such as adaptations and homologous structures, which is one of the main criterion scientists use to evaluate the merits of a scientific theory. However, like all theories in science, the principles of descent with modification must be tested in many different ways before it can be considered valid or acceptable by the scientific community. In this investigation, you will therefore test this theory by examining the process of embryo development in animals to determine if it is consistent with the major principles of descent with modification.

Reference

Darwin, C. 1845. *Journal of researches into the natural history and geology of the countries visited during the voyage of H.M.S. Beagle round the world, under the Command of Capt. Fitz Roy, R.N.* 2nd ed. London: John Murray.

Your Task

Collect data about the embryonic development of eight different animals. Then use the data you collect to test the theory of descent with modification.

The guiding question of this investigation is, **Does animal embryonic development support or refute the theory of descent with modification?**

Materials

Your teacher will provide you with a set of the following images that can be used during your investigation:

- Images of amphibian (frog) embryo development
- Images of bird (fieldfare) embryo development
- Images of bird (quail) embryo development
- Images of fish (zebrafish) embryo development
- Images of mammal (bat) embryo development
- Images of mammal (mouse) embryo development
- Images of reptile (snake) embryo development
- Images of reptile (turtle) embryo development

Safety Precautions

Follow all normal lab safety rules.

Investigation Proposal Required?   ☐ Yes   ☐ No

Getting Started

All animals begin life as a single cell. This single cell then starts to divide and becomes an embryo. As the cells continue to divide, specific structures such as organs and limbs begin to take shape. The embryo eventually starts to look more and more like the animal it will become as it changes and grows in size. Biologists have been studying the process of embryo development in animals since the 19th century. We therefore know a lot about what animal embryos look like as they develop and grow over time.

To answer the guiding question, you will need to compare and contrast the process of embryo development for eight different animals. You will be supplied with 10 embryo images for each animal. The images represent specific embryonic development milestones. You must determine what type of data you need to collect from these images, how you will collect it, and how you will analyze it. To determine *what type of data you need to collect*, think about the following questions:

- What would you expect the process of embryo development to look like in these eight different animals if they shared a common ancestor? What would it look like if they did not share a common ancestor?
- Which animals are more closely related to each other?
- Which characteristics of the embryo will you examine?
- How many different characteristics of the embryos will you need to examine?

To determine *how you will collect the data*, think about the following questions:

- How will you quantify differences and similarities in embryos?
- How will you make sure that your data are of high quality?
- What will you do with the data you collect?

To determine *how you will analyze your data*, think about the following questions:

- How will you compare and contrast the various embryos?
- What type of graph or table could you create to help make sense of your data?

LAB 20

Connections to Crosscutting Concepts, the Nature of Science, and the Nature of Scientific Inquiry

As you work through your investigation, be sure to think about

- the importance of looking for patterns in nature,
- the relationship between structure and function in nature,
- how science as a body of knowledge develops over time, and
- the different types of methods that scientists use to answer questions.

Initial Argument

Once your group has finished collecting and analyzing your data, you will need to develop an initial argument. Your argument must include a claim, evidence to support your claim, and a justification of the evidence. The claim is your group's answer to the guiding question. The evidence is an analysis and interpretation of your data. Finally, the justification of the evidence is why your group thinks the evidence matters. The justification of the evidence is important because scientists can use different kinds of evidence to support their claims. Your group will create your initial argument on a whiteboard. Your whiteboard should include all the information shown in Figure L20.3.

FIGURE L20.3

Argument presentation on a whiteboard

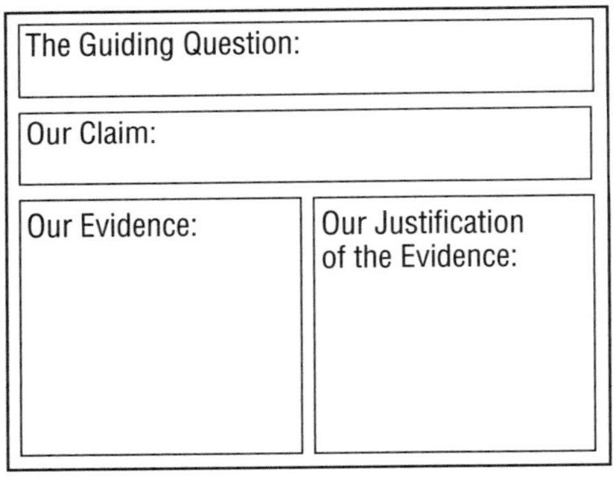

The Guiding Question:	
Our Claim:	
Our Evidence:	Our Justification of the Evidence:

Argumentation Session

The argumentation session allows all of the groups to share their arguments. One member of each group will stay at the lab station to share that group's argument, while the other members of the group go to the other lab stations one at a time to listen to and critique the arguments developed by their classmates. This is similar to how scientists present their arguments to other scientists at conferences. If you are responsible for critiquing your classmates' arguments, your goal is to look for mistakes so these mistakes can be fixed and they can make their argument better. The argumentation session is also a good time to think about ways you can make your initial argument better. Scientists must share and critique arguments like this to develop new ideas.

To critique an argument, you might need more information than what is included on the whiteboard. You will therefore need to ask the presenter lots of questions. Here are some good questions to ask:

- What did your group do to analyze the data? Why did your group decide to analyze it that way?
- What other ways of analyzing and interpreting the data did your group talk about?

- Why did your group decide to present your evidence in that way?

- Why did your group abandon the other explanations?

- How sure are you that your group's claim is accurate? What could you do to be more certain?

Once the argumentation session is complete, you will have a chance to meet with your group and revise your initial argument. Your group might need to gather more data as part of this process. Remember, your goal at this stage of the investigation is to develop the best argument possible.

Report

Once you have completed your research, you will need to prepare an investigation report that consists of three sections that provide answers to the following questions:

1. What question were you trying to answer and why?

2. What did you do during your investigation and why did you conduct your investigation in this way?

3. What is your argument?

Your report should answer these questions in two pages or less. The report must be typed, and any diagrams, figures, or tables should be embedded into the document. Be sure to write in a persuasive style; you are trying to convince others that your claim is acceptable or valid!

LAB 20

Lab 20. Descent With Modification and Embryonic Development: Does Animal Embryonic Development Support or Refute the Theory of Descent With Modification?

1. What are the basic principles of the theory of descent with modification?

2. Use the theory of descent with modification to explain why all mammals have the same set of bones in their limbs. The forelimbs of dogs and whales, for example, include a humerus, an ulna, a radius, and several carpals.

3. Scientific knowledge changes and develops over time.

   a. I agree with this statement.

   b. I disagree with this statement.

   Explain your answer, using an example from your investigation about descent with modification and embryonic development.

4. In science, there are usually multiple ways to investigate a question.

   a. I agree with this statement.

   b. I disagree with this statement.

   Explain your answer, using an example from your investigation about descent with modification and embryonic development.

5. Scientists often look for patterns during their investigations. Explain why patterns are important to look for, using an example from your investigation about descent with modification and embryonic development.

6. The relationship between the structure and function of organisms' features is an important area of study in science. Discuss why it is important for scientists to understand this relationship, using an example from your investigation about descent with modification and embryonic development.

SECTION 6
Appendixes

APPENDIX 1
Standards Alignment Matrixes

Alignment of the Argument-Driven Inquiry Lab Investigations With the Scientific Practices, Crosscutting Concepts, and Core Ideas in *A Framework for K–12 Science Education* (NRC 2012)

Aspect of the NRC *Framework*	1-Cellular Respiration	2-Photosynthesis	3-Osmosis	4-Cell Structure	5-Temperature and Photosynthesis	6-Energy in Food	7-Respiratory and Cardiovascular Systems	8-Memory and Stimuli	9-Population Growth	10-Predator-Prey Relationships	11-Food Webs and Ecosystems	12-Matter in Ecosystems	13-Carbon Cycling	14-Variation in Traits	15-Mutations in Genes	16-Mechanisms of Inheritance	17-Mechanisms of Evolution	18-Environmental Change and Evolution	19-Phylogenetic Trees and the Classification of Fossils	20-Descent With Modification and Embryonic Development
Scientific practices																				
Asking questions and defining problems	□	□	□	□	□	□	□	□	□	□	□	□	□	□	□	□	□	□	□	□
Developing and using models	■	■	■	■	■				■	■	■	■	■		■	■	■	■		
Planning and carrying out investigations	■	■	■	■	■	■	■	■	■	■	■	■	■	■	■	■	■	■	■	■
Analyzing and interpreting data	■	■	■	■	■	■	■	■	■	■	■	■	■	■	■	■	■	■	■	■
Using mathematics and computational thinking			■		■		■			■		■	■				■	■	■	
Constructing explanations and designing solutions	■	■		■	■	■	■			■	■	■	■	■	■	■	■	■	■	■
Engaging in argument from evidence	■	■	■	■	■	■	■	■	■	■	■	■	■	■	■	■	■	■	■	■
Obtaining, evaluating, and communicating information	■	■	■	■	■	■	■	■	■	■	■	■	■	■	■	■	■	■	■	■
Crosscutting concepts																				
Patterns				■				■	■					■		■	■		■	■
Cause and effect: Mechanism and explanation	■				■					■							■			

Key: ■ = strong alignment; □ = moderate alignment

Alignment of the Argument-Driven Inquiry Lab Investigations With the Scientific Practices, Crosscutting Concepts, and Core Ideas in *A Framework for K–12 Science Education* (NRC 2012) *(continued)*

Aspect of the NRC *Framework*	Lab Investigation																			
	1-Cellular Respiration	2-Photosynthesis	3-Osmosis	4-Cell Structure	5-Temperature and Photosynthesis	6-Energy in Food	7-Respiratory and Cardiovascular Systems	8-Memory and Stimuli	9-Population Growth	10-Predator-Prey Relationships	11-Food Webs and Ecosystems	12-Matter in Ecosystems	13-Carbon Cycling	14-Variation in Traits	15-Mutations in Genes	16-Mechanisms of Inheritance	17-Mechanisms of Evolution	18-Environmental Change and Evolution	19-Phylogenetic Trees and the Classification of Fossils	20-Descent With Modification and Embryonic Development
Scale, proportion, and quantity		■											■		■					
Systems and system models			■				■			■	■					■	■	■		
Energy and matter: Flows, cycles, and conservation	■	■	■		■	■						■	■							
Structure and function				■		■	■	■						■	■			■	■	■
Stability and change									■			■	■							
Core ideas																				
LS1: From molecules to organisms: Structures and processes	■	■	■	■	■	■	■	■												
LS2: Ecosystems: Interactions, energy, and dynamics									■	■	■	■	■				■	■		
LS3: Heredity: Inheritance and variation of traits														■	■	■	■	■		
LS4: Biological evolution: Unity and diversity																	■	■	■	■

Key: ■ = strong alignment; □ = moderate alignment

Alignment (■) of the Argument-Driven Inquiry Lab Investigations With the *Common Core State Standards* for English Language Arts and Mathematics (NGAC and CCSSO 2010)

Standard	\[Lab Investigation\] 1-Cellular Respiration	2-Photosynthesis	3-Osmosis	4-Cell Structure	5-Temperature and Photosynthesis	6-Energy in Food	7-Respiratory and Cardiovascular Systems	8-Memory and Stimuli	9-Population Growth	10-Predator-Prey Relationships	11-Food Webs and Ecosystems	12-Matter in Ecosystems	13-Carbon Cycling	14-Variation in Traits	15-Mutations in Genes	16-Mechanisms of Inheritance	17-Mechanisms of Evolution	18-Environmental Change and Evolution	19-Phylogenetic Trees and the Classification of Fossils	20-Descent With Modification and Embryonic Development
Reading																				
Key ideas and details	■	■	■	■	■	■	■	■	■	■	■	■	■	■	■	■	■	■	■	■
Craft and structure	■	■	■	■	■	■	■	■	■	■	■	■	■	■	■	■	■	■	■	■
Integration of knowledge and ideas	■	■	■	■	■	■	■	■	■	■	■	■	■	■	■	■	■	■	■	■
Writing																				
Text types and purposes	■	■	■	■	■	■	■	■	■	■	■	■	■	■	■	■	■	■	■	■
Production and distribution of writing	■	■	■	■	■	■	■	■	■	■	■	■	■	■	■	■	■	■	■	■
Research to build and present knowledge	■	■	■	■	■	■	■	■	■	■	■	■	■	■	■	■	■	■	■	■
Range of writing	■	■	■	■	■	■	■	■	■	■	■	■	■	■	■	■	■	■	■	■
Speaking and listening																				
Comprehension and collaboration	■	■	■	■	■	■	■	■	■	■	■	■	■	■	■	■	■	■	■	■

Key: ■ = strong alignment; □ = moderate alignment

Alignment (■) of the Argument-Driven Inquiry Lab Investigations With the *Common Core State Standards* for English Language Arts and Mathematics (NGAC and CCSSO 2010) *(continued)*

Standard	Lab Investigation																			
	1-Cellular Respiration	2-Photosynthesis	3-Osmosis	4-Cell Structure	5-Temperature and Photosynthesis	6-Energy in Food	7-Respiratory and Cardiovascular Systems	8-Memory and Stimuli	9-Population Growth	10-Predator-Prey Relationships	11-Food Webs and Ecosystems	12-Matter in Ecosystems	13-Carbon Cycling	14-Variation in Traits	15-Mutations in Genes	16-Mechanisms of Inheritance	17-Mechanisms of Evolution	18-Environmental Change and Evolution	19-Phylogenetic Trees and the Classification of Fossils	20-Descent With Modification and Embryonic Development
Presentation of knowledge and ideas	■	□	■	■	■	■	■	■	■	■	■	■	■	■	■	■	■	■	■	■
Mathematics																				
Make sense of problems and persevere in solving them	■	■	■	■	■	■	■	■	■	■	■	■	■	■	■	■	■	■	■	■
Reason abstractly and quantitatively	■	■	■		■	■	■	■	■	■		■	■		■	■	■	■	■	■
Construct viable arguments and critique the reasoning of others	■	■	■	■	■	■	■	■	■	■	■	■	■	■	■	■	■	■	■	■
Model with mathematics					■	■	■		■	■		■	■					■	■	
Use appropriate tools strategically	■	■	■		■	■	■		■	■	■	■	■		■	■	■	■	■	■
Attend to precision					■	■														
Look for and make use of structure					■	■										■				
Look for and express regularity in repeated reasoning					■	■		■	■				■				■		■	■

Key: ■ = strong alignment; □ = moderate alignment

Alignment (■) of the Argument-Driven Inquiry Lab Investigations With the Nature of Science (NOS) and the Nature of Scientific Inquiry (NOSI) Concepts*

NOS or NOSI concept*	1-Cellular Respiration	2-Photosynthesis	3-Osmosis	4-Cell Structure	5-Temperature and Photosynthesis	6-Energy in Food	7-Respiratory and Cardiovascular Systems	8-Memory and Stimuli	9-Population Growth	10-Predator-Prey Relationships	11-Food Webs and Ecosystems	12-Matter in Ecosystems	13-Carbon Cycling	14-Variation in Traits	15-Mutations in Genes	16-Mechanisms of Inheritance	17-Mechanisms of Evolution	18-Environmental Change and Evolution	19-Phylogenetic Trees and the Classification of Fossils	20-Descent With Modification and Embryonic Development
Observations and inferences	■		■	■					■			■		■					■	
Changes in scientific knowledge over time				■						■				■					■	■
Scientific laws and theories					■										■	■	■			
Social and cultural influences						■	■	■			■									
Difference between data and evidence	■				■							■	■			■				
Methods used in scientific investigations		■								■			■				■	■		■
Imagination and creativity in science						■		■			■				■		■			
Nature and role of experiments		■	■				■		■											

Key: ■ = strong alignment; □ = moderate alignment

*The NOS/NOSI concepts listed in this matrix are based on the work of Abd-El-Khalick and Lederman 2000; Akerson, Abd-El-Khalick, and Lederman 2000; Lederman et al. 2002, 2014; and Schwartz, Lederman, and Crawford 2004.

References

Abd-El-Khalick, F., and N. G. Lederman. 2000. Improving science teachers' conceptions of nature of science: A critical review of the literature. *International Journal of Science Education* 22: 665–701.

Akerson, V., F. Abd-El-Khalick, and N. Lederman. 2000. Influence of a reflective explicit activity-based approach on elementary teachers' conception of nature of science. *Journal of Research in Science Teaching* 37 (4): 295–317.

Lederman, N. G., F. Abd-El-Khalick, R. L. Bell, and R. S. Schwartz. 2002. Views of nature of science questionnaire: Toward a valid and meaningful assessment of learners' conceptions of nature of science. *Journal of Research in Science Teaching* 39 (6): 497–521.

Lederman, J., N. Lederman, S. Bartos, S. Bartels, A. Meyer, and R. Schwartz. 2014. Meaningful assessment of learners' understanding about scientific inquiry: The Views About Scientific Inquiry (VASI) questionnaire. *Journal of Research in Science Teaching* 51 (1): 65–83.

National Governors Association Center for Best Practices and Council of Chief State School Officers (NGAC and CCSSO). 2010. *Common core state standards.* Washington, DC: NGAC and CCSSO.

National Research Council (NRC). 2012. *A framework for K–12 science education: Practices, crosscutting concepts, and core ideas.* Washington, DC: National Academies Press.

Schwartz, R. S., N. Lederman, and B. Crawford. 2004. Developing views of nature of science in an authentic context: An explicit approach to bridging the gap between nature of science and scientific inquiry. *Science Education* 88: 610–645.

APPENDIX 2
Options for Implementing ADI Lab Investigations

Note: Compliance with safety precautions should be addressed during stages 1 and 2.

Option A

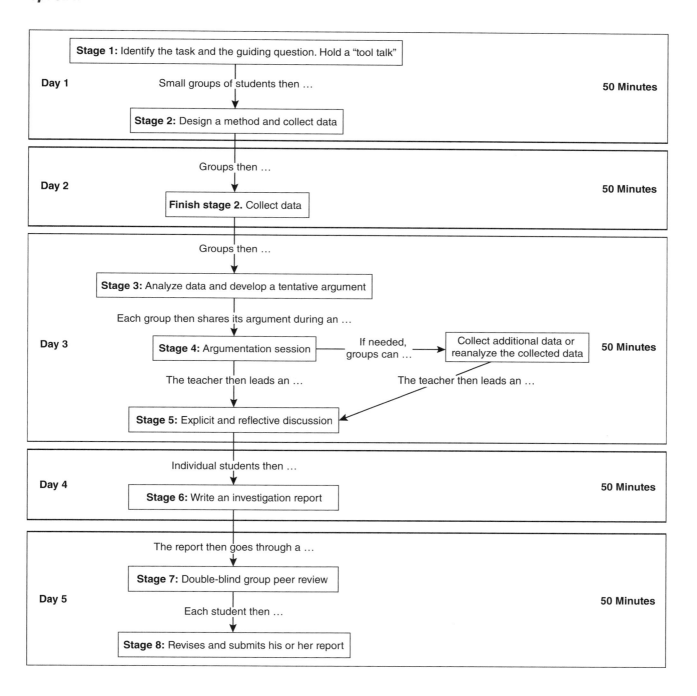

Option B

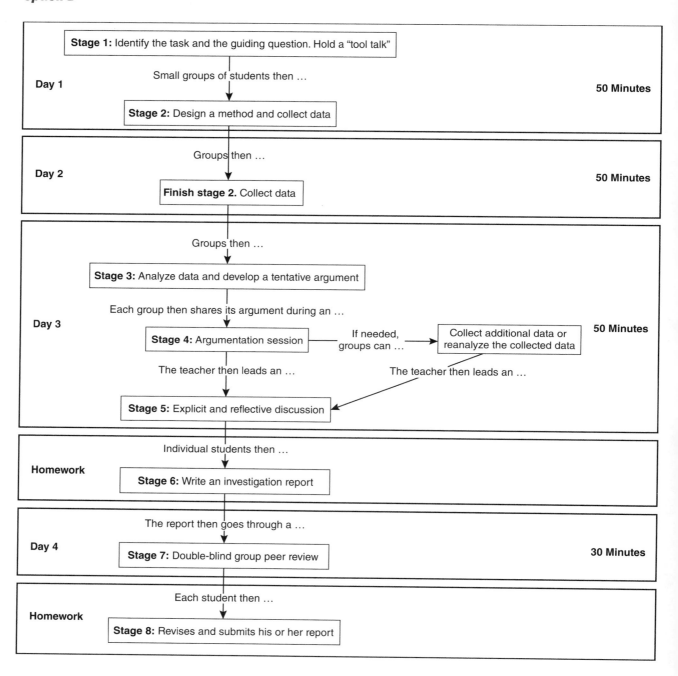

Option C

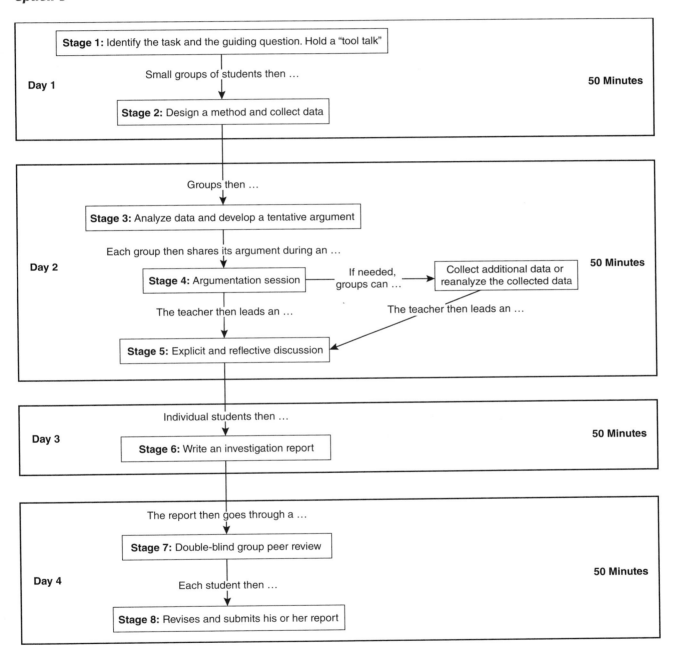

Stage 1: Identify the task and the guiding question. Hold a "tool talk"

Day 1 Small groups of students then … **50 Minutes**

Stage 2: Design a method and collect data

Groups then …

Stage 3: Analyze data and develop a tentative argument

Each group then shares its argument during an …

Day 2 Stage 4: Argumentation session If needed, groups can … → Collect additional data or reanalyze the collected data **50 Minutes**

The teacher then leads an … The teacher then leads an …

Stage 5: Explicit and reflective discussion

Individual students then …

Day 3 Stage 6: Write an investigation report **50 Minutes**

The report then goes through a …

Stage 7: Double-blind group peer review

Day 4 Each student then … **50 Minutes**

Stage 8: Revises and submits his or her report

Option D

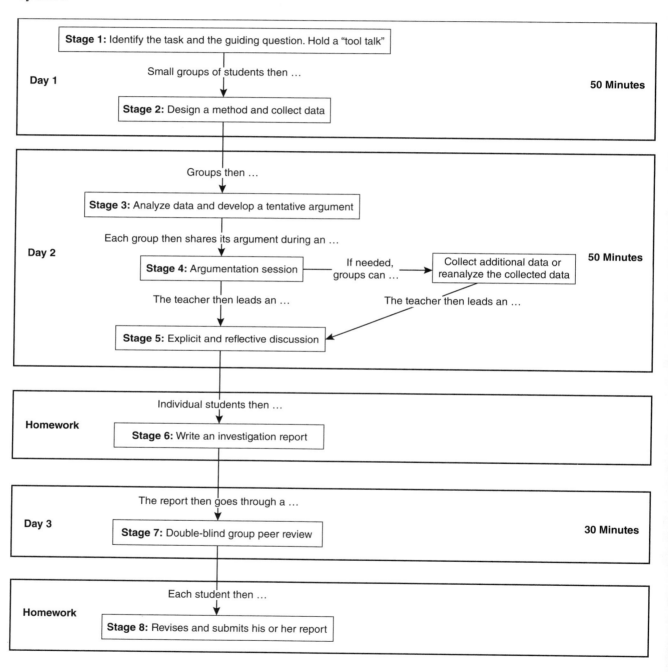

Option E

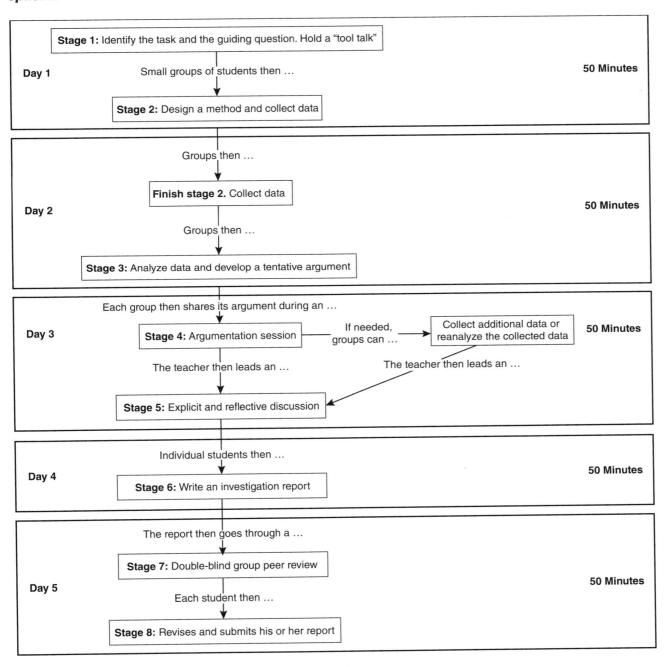

Option F

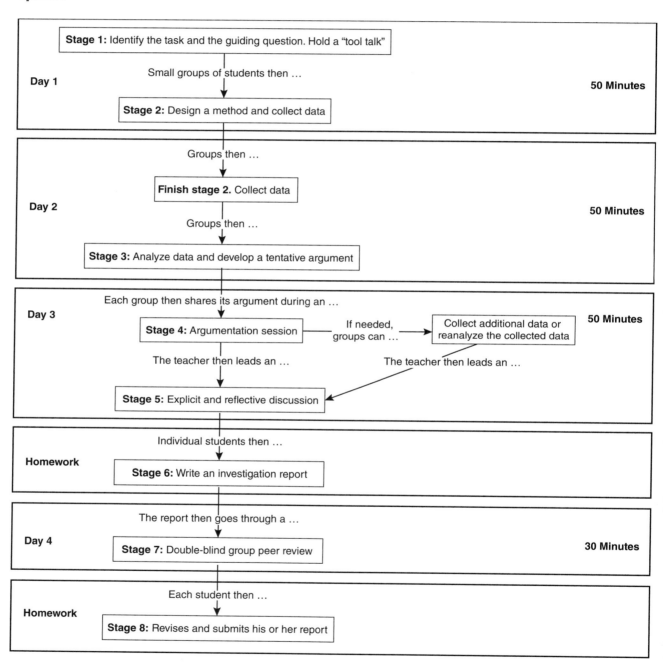

APPENDIX 3

Investigation Proposal Options

This appendix presents three investigation proposals that may be used in most labs. The development of these proposals was supported by the Institute of Education Sciences, U.S. Department of Education, through grant R305A100909 to Florida State University.

The format of investigation proposals A and B is modeled after a hypothetical deductive-reasoning guide described in *Exploring the Living World* (Lawson 1995) and modified from an investigation guide described in an article by Maguire, Myerowitz, and Sampson (2010).

References

Lawson, A. E. 1995. *Exploring the living world: A laboratory manual for biology.* McGraw-Hill College.

Maguire, L., L. Myerowitz, and V. Sampson. 2010. Diffusion and osmosis in cells: A guided inquiry activity. *The Science Teacher* 77 (8): 55–60.

Investigation Proposal A

The Guiding Question ... [   ]

Hypothesis 1 ← → Hypothesis 2

IF ... [   ]   IF ... [   ]

The Test

AND ...
Procedure

What data will you collect?

How will you analyze the data?

What safety precautions will you follow?

Predicted Result if hypothesis 1 is valid   **Predicted Result** if hypothesis 2 is valid

THEN ... [   ]   THEN ... [   ]

The Actual Results   AND ... [   ]

I approve of this investigation. _____   _____

Instructor's Signature   Date

The development of this investigation proposal was supported by the Institute of Education Sciences, U.S. Department of Education, through Grant R305A100909 to the Florida State University. The format of the proposal is modeled after a hypothetical deductive-reasoning guide described in *Exploring the Living World* (Lawson 1995) and modified from an investigation guide described in Macquire, Myerowitz, and Sampson (2010).

Investigation Proposal B

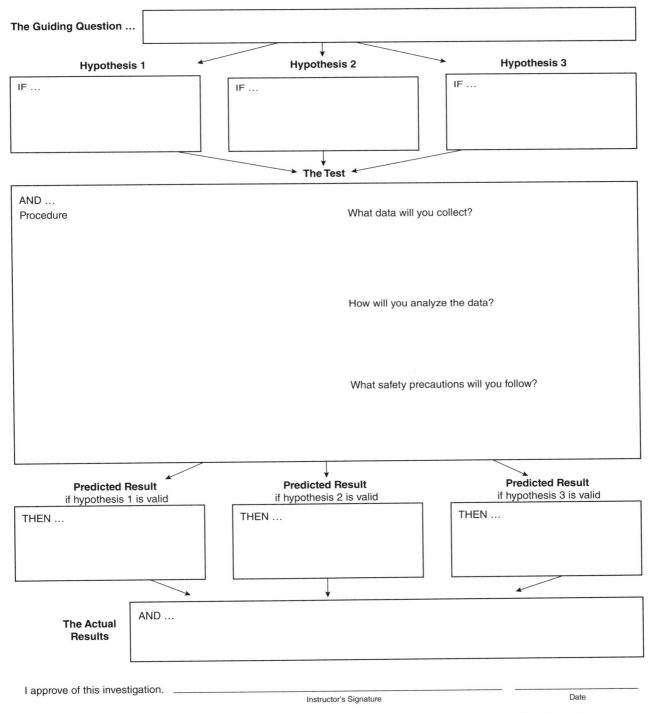

The Guiding Question ...

Hypothesis 1

IF ...

Hypothesis 2

IF ...

Hypothesis 3

IF ...

The Test

AND ...
Procedure

What data will you collect?

How will you analyze the data?

What safety precautions will you follow?

Predicted Result
if hypothesis 1 is valid

THEN ...

Predicted Result
if hypothesis 2 is valid

THEN ...

Predicted Result
if hypothesis 3 is valid

THEN ...

The Actual Results

AND ...

I approve of this investigation. _____   _____

Instructor's Signature   Date

The development of this investigation proposal was supported by the Institute of Education Sciences, U.S. Department of Education, through Grant R305A100909 to the Florida State University. The format of the proposal is modeled after a hypothetical deductive-reasoning guide described in *Exploring the Living World* (Lawson 1995) and modified from an investigation guide described in Macquire, Myerowitz, and Sampson (2010).

Investigation Proposal C

The Guiding Question ...	

↓

What data will you collect?	

↓

How will you collect your data?	Your Procedure What safety precautions will you follow?

↓

How will you analyze your data?	

↓

Your actual data	

I approve of this investigation. _____ _____

                                       Instructor's Signature   Date

The development of this investigation proposal was supported by the Institute of Education Sciences, U.S. Department of Education, through Grant R305A100909 to the Florida State University.

APPENDIX 4

Investigaton Report Peer-Review Guide: Middle School Version

Report By: _____
ID Number

Reviewed By: _____   _____   _____   _____
ID Number   ID Number   ID Number   ID Number

Author: Did the reviewers do a good job?   1 2 3 4 5

Rate the overall quality of the peer review

Section 1: Introduction and Guiding Question	Reviewer Rating			Instructor Score
1. Did the author provide enough **background information**?	☐ No	☐ Partially	☐ Yes	0 1 2
2. Is the background information **correct**?	☐ No	☐ Partially	☐ Yes	0 1 2
3. Did the author make the **goal of the investigation** clear?	☐ No	☐ Partially	☐ Yes	0 1 2
4. Did the author make the **guiding question** clear?	☐ No	☐ Partially	☐ Yes	0 1 2

Reviewers: If your group made any "No" or "Partially" marks in this section, please explain how the author could improve this part of his or her report.

Author: What revisions did you make in your report? Is there anything you decided to keep the same even though the reviewers suggested otherwise? Be sure to explain why.

Section 2: Method	Reviewer Rating			Instructor Score
1. Did the author provide a clear description of what he or she did during the investigation to **collect data** (the method)?	☐ No	☐ Partially	☐ Yes	0 1 2
2. Did the author describe **how** he or she **analyzed** the data?	☐ No	☐ Partially	☐ Yes	0 1 2
3. Did the author use the **correct term** to describe his or her investigation (e.g., experiment, observations, interpretation of a data set)?	☐ No	☐ Partially	☐ Yes	0 1 2

Section 2: Method *(continued)*	Reviewer Rating	Instructor Score
Reviewers: If your group made any "No" or "Partially" marks in this section, please explain how the author could improve this part of his or her report.	**Author:** What revisions did you make in your report? Is there anything you decided to keep the same even though the reviewers suggested otherwise? Be sure to explain why.	

Section 3: The Argument	Reviewer Rating			Instructor Score
1. Did the author provide a **clear and complete claim** that answers the guiding question?	☐ No	☐ Partially	☐ Yes	0 1 2
2. Did the author use **evidence** to support his or her claim? Evidence is an analysis of data and an explanation of what the analysis means.	☐ No	☐ Partially	☐ Yes	0 1 2
	☐ No	☐ Partially	☐ Yes	0 1 2
	☐ No	☐ Partially	☐ Yes	0 1 2
3. Did the author **present the evidence** in an appropriate manner by • including a correctly formatted and labeled graph (or table); • using correct metric units (e.g., m/s, g, ml); and • referencing the graph or table in the body of the text?	☐ No	☐ Partially	☐ Yes	0 1 2
	☐ No	☐ Partially	☐ Yes	0 1 2
	☐ No	☐ Partially	☐ Yes	0 1 2
4. Is the **evidence support the author's claim**?	☐ No	☐ Partially	☐ Yes	0 1 2
5. Did the author use a scientific concept to **justify the evidence**? The justification of the evidence explains why the evidence matters.	☐ No	☐ Partially	☐ Yes	0 1 2
	☐ No	☐ Partially	☐ Yes	0 1 2
6. Is the **justification of the evidence** acceptable?	☐ No	☐ Partially	☐ Yes	0 1 2
7. Did the author **use scientific terms correctly** (e.g., *hypothesis* vs. *prediction*, *data* vs. *evidence*) and **reference the evidence in an appropriate manner** (e.g., *supports* or *suggests* vs. *proves*)?	☐ No	☐ Partially	☐ Yes	0 1 2

Section 3: The Argument *(continued)*	Reviewer Rating	Instructor Score
Reviewers: If your group made any "No" or "Partially" marks in this section, please explain how the author could improve this part of his or her report.	**Author:** What revisions did you make in your report? Is there anything you decided to keep the same even though the reviewers suggested otherwise? Be sure to explain why.	

Mechanics	Reviewer Rating			Instructor Score
1. *Organization:* Is each section easy to follow? Do paragraphs include multiple sentences? Do paragraphs begin with a topic sentence?	☐ No	☐ Partially	☐ Yes	0 1 2
2. *Grammar:* Are the sentences complete? Is there proper subject-verb agreement in each sentence? Are there run-on sentences?	☐ No	☐ Partially	☐ Yes	0 1 2
3. *Conventions:* Did the author use appropriate spelling, punctuation, and capitalization?	☐ No	☐ Partially	☐ Yes	0 1 2
4. *Word Choice:* Did the author use the appropriate word (e.g., *there* vs. *their*, *to* vs. *too*, *than* vs. *then*)?	☐ No	☐ Partially	☐ Yes	0 1 2

Instructor Comments:

Total: _____ /40

IMAGE CREDITS

CHAPTER 1

Figure 4: Authors

Figure 5: Authors

Lab 1

Figure 1.1: OpenStax College, Wikimedia Commons, CC BY 3.0. *http://commons.wikimedia.org/wiki/File:0315_Mitochondrion_new.jpg*

Figure 1.2: Adapted from User:Boumphreyfr, Wikimedia Commons, CC BY-SA 3.0. *http://commons.wikimedia.org/wiki/File:Aerobic_mitochondria_process.png*

Figure L1.1: OpenStax College, Wikimedia Commons, CC BY 3.0. *http://commons.wikimedia.org/wiki/File:0315_Mitochondrion_new.jpg*

Figure L1.2: Courtesy of Jonathon Grooms.

Figure L1.3: Authors

Lab 2

Figure 2.1*:* Daniel Mayer, Wikimedia Commons, CC BY-SA 4.0, GFDL 1.2. *http://commons.wikimedia.org/wiki/File:Simple_photosynthesis_overview.svg*

Figure 2.2: User:Ollin, Wikimedia Commons, Public domain. *http://commons.wikimedia.org/wiki/File:Chloroplast_diagram.svg*

Figure L2.1: User:At09kg, Wikimedia Commons, CC BY-SA 3.0. *http://commons.wikimedia.org/wiki/File:Plants.gif*

Figure L2.2: Authors

Figure L2.3: Authors

Checkout Questions figure: Authors

Lab 3

Figure 3.1: User:Pidalka44, Wikimedia Commons, Public domain. *http://commons.wikimedia.org/wiki/File:Semipermeable_membrane.png*

Figure 3.2: User:LadyofHats, Wikimedia Commons, Public domain. *http://commons.wikimedia.org/wiki/File:Osmotic_pressure_on_blood_cells_diagram-sk.svg*

Figure L3.1: User:Pidalka44, Wikimedia Commons, Public domain. *http://commons.wikimedia.org/wiki/File:Semipermeable_membrane.png*

Figure L3.2: User:LadyofHats, Wikimedia Commons, Public domain. *http://commons.wikimedia.org/wiki/File:Osmotic_pressure_on_blood_cells_diagram-sk.svg*

Figure L3.3: Courtesy of Jonathon Grooms.

Figure L3.4: Authors

Checkout Questions figures: Authors

Lab 4

Figure 4.1: OpenStax College, Wikimedia Commons, CC BY 3.0. *http://commons.wikimedia.org/wiki/File:0312_Animal_Cell_and_Components.jpg*

Figure L4.1: OpenStax College, Wikimedia Commons, CC BY 3.0. *http://commons.wikimedia.org/wiki/File:0312_Animal_Cell_and_Components.jpg*

Figure L4.2: Authors

Lab 5

Figure 5.1: Daniel Mayer, Wikimedia Commons, CC BY-SA 4.0, GFDL 1.2. *http://commons.wikimedia.org/wiki/File:Simple_photosynthesis_overview.svg*

Figure 5.2: Adapted from Mikael Häggström, Wikimedia Commons, CC BY-SA 3.0, GFDL 1.2. *http://commons.wikimedia.org/wiki/File:Auto-and_heterotrophs.svg*

Figure L5.1: Daniel Mayer, Wikimedia Commons, CC BY-SA 4.0, GFDL 1.2. *http://commons.wikimedia.org/wiki/File:Simple_photosynthesis_overview.svg*

Figure L5.2: Authors

Figure L5.3: Authors

Lab 6

Figure 6.1: Courtesy of Jonathon Grooms.

Figure L6.1: U.S. Food and Drug Administration, Wikimedia Commons, Public domain. *http://commons.wikimedia.org/wiki/File%3AFDA_Nutrition_Facts_Label_2006.jpg*

Figure L6.2: Courtesy of Jonathon Grooms.

Figure L6.3: Authors

Lab 7

Figure 7.1: Theresa Knott, Wikimedia Commons, CC BY-SA 2.5, GFDL 1.2. *http://commons.wikimedia.org/wiki/File:Respiratory_system.svg*

Figure 7.2: User:Sansculotte, Wikimedia Commons, CC BY-SA 3.0. *http://commons.wikimedia.org/wiki/File:Grafik_blutkreislauf.jpg*

Figure L7.1: Theresa Knott, Wikimedia Commons, CC BY-SA 2.5, GFDL 1.2. *http://commons.wikimedia.org/wiki/File:Respiratory_system.svg*

Figure L7.2: User:Sansculotte, Wikimedia Commons, CC BY-SA 3.0. *http://commons.wikimedia.org/wiki/File:Grafik_blutkreislauf.jpg*

Figure L7.3: Authors

Lab 8

Figure L8.1: Authors

Checkout Questions figure: Authors

Lab 9

Figure 9.1: Gerhard K. Heilig, Wikimedia Commons, CC BY-SA 3.0. *http://commons.wikimedia.org/wiki/File:Rapid-Population-Growth_Rev5.png*

Figure L9.1: Bob Blaylock, Wikimedia Commons, CC BY-SA 3.0. *http://commons.wikimedia.org/wiki/File:20100911_232323_Yeast_Live.jpg*

Figure L9.2: Authors

Lab 10

Figure 10.1: User:Praguepower, Wikimedia Commons, CC BY-SA 3.0. *http://commons.wikimedia.org/wiki/File:Predator_prey.jpg*

Figure L10.1: Authors

Lab 11

Figure 11.1: Tyler Rubley, Wikimedia Commons, CC BY-SA 3.0. *http://commons.wikimedia.org/wiki/File:Mosquito_Energy_Transfer_Food_Web.pdf*

Figure L11.1: Tyler Rubley, Wikimedia Commons, CC BY-SA 3.0. *http://commons.wikimedia.org/wiki/File:Mosquito_Energy_Transfer_Food_Web.pdf*

Figure L11.2: Authors

Checkout Questions figure: Authors

Lab 12

Figure 12.1: Harry C, Wikimedia Commons, CC BY-SA 3.0. *http://commons.wikimedia.org/wiki/File:Carbon-cycle-full.jpg*

Figure L12.1: Authors

Lab 12 Reference Sheet

Nitrogen cycle: Johann Dréo, Wikimedia Commons, CC BY-SA 3.0, GFDL 1.2. *http://commons.wikimedia.org/wiki/File:Cicle_del_nitrogen_ca.svg*

Phosphorus cycle: User:Bonniemf, Wikimedia Commons, CC BY-SA 3.0. *http://commons.wikimedia.org/wiki/File:Phosphorus_cycle.png*

Lab 13

Figure 13.1: Adapted from Harry C, Wikimedia Commons, CC BY-SA 3.0. *http://commons.wikimedia.org/wiki/File:Carbon-cycle-full.jpg*

Figure L13.1: Adapted from Harry C, Wikimedia Commons, CC BY-SA 3.0. *http://commons.wikimedia.org/wiki/File:Carbon-cycle-full.jpg*

Figure L13.2: Authors

Lab 14

Figure 14.1: User:Pengo, Wikimedia Commons, Public domain. *http://commons.wikimedia.org/wiki/File:Biological_classification_L_Pengo_tweaked.svg*

Figure 14.2: Blatchley, W. S. 1859-1940, Wikimedia Commons, Public domain. *http://commons.wikimedia.org/wiki/File:Dynastes_tityusBlatchleyF312A.jpg*

Figure 14.3: User:Siga, Wikimedia Commons, CC BY-SA 3.0, GFDL 1.2 *http://commons.wikimedia.org/wiki/File:Carabus_violaceus_up.jpg*

Figure L14.1: Blatchley, W. S. 1859-1940, Wikimedia Commons, Public domain. *http://commons.wikimedia.org/wiki/File:Dynastes_tityusBlatchleyF312A.jpg*

Figure L14.2: User:Siga, Wikimedia Commons, CC BY-SA 3.0, GFDL 1.2. *http://commons.wikimedia.org/wiki/File:Carabus_violaceus_up.jpg*

Figure L14.3: Authors

Lab 14 Reference Sheet

Harpalus affinis: a: User:©entomart, Wikimedia Commons. *http://commons.wikimedia.org/wiki/File:Harpalus_affinis01.jpg; b:* James Lindsey at Ecology of Commanster, Wikimedia Commons, CC BY-SA 3.0. *http://commons.wikimedia.org/wiki/File:Harpalus.affinis.jpg; c:* User:Futureman1199, Wikimedia Commons, CC BY-SA 3.0. *http://commons.wikimedia.org/wiki/File:Harpalus_affinis.jpg*

Cotinis mutabilis: a: User:Davefoc, Wikimedia Commons, CC BY-SA 3.0, GFDL 1.2. *http://commons.wikimedia.org/wiki/File:CotinisMutabilis_7871.JPG;* b: User:Davefoc, Wikimedia Commons, CC BY-SA 3.0, GFDL 1.2. *http://commons.wikimedia.org/wiki/File:CotinisMutabilis_7864.JPG;* c: Eugene Zelenko, Wikimedia Commons, CC BY-SA 3.0, GFDL 1.2. *http://commons.wikimedia.org/wiki/File:Cotinis_mutabilis-3.jpg.*

Leptinotarsa decemlineata: a: Fritz Geller-Grimm, Wikimedia Commons, CC BY-SA 3.0. *http://commons.wikimedia.org/wiki/File:Leptinotarsa_fg02.jpg;* b: User:Barbarossa, Wikimedia Commons, CC BY-SA 3.0, GFDL 1.2. *http://commons.wikimedia.org/wiki/File:Coloradokever_dichtbij.png;* c: Scott Bauer, USDA ARS, Wikimedia Commons, Public domain. *http://commons.wikimedia.org/wiki/File:Colorado_potato_beetle.jpg.*

Lab 15

Figure 15.1: Adapted from すじにくシチュー, Wikimedia Commons, Public domain. *http://commons.wikimedia.org/wiki/File:DNA%E3%81%AE%E4%B8%A6%E3%81%B3%E6%96%B9.png*

Figure 15.2: Jessica Reuter, Wikimedia Commons, Public domain. *http://commons.wikimedia.org/wiki/File:Central_dogma.JPG*

Figure 15.3: Courtesy of Patrick Enderle.

Figure L15.1: Adapted from すじにくシチュー, Wikimedia Commons, Public domain. *http://commons.wikimedia.org/wiki/File:DNA%E3%81%AE%E4%B8%A6%E3%81%B3%E6%96%B9.png*

Figure L15.2: Jessica Reuter, Wikimedia Commons, Public domain. *http://commons.wikimedia.org/wiki/File:Central_dogma.JPG*

Figure L15.3: Courtesy of Patrick Enderle.

Figure L15.4: Authors

Lab 16

Figure L16.1: User:Madboy74, Wikimedia Commons, Public domain. *http://commons.wikimedia.org/wiki/File:Biology_Illustration_Animals_Insects_Drosophila_melanogaster.svg*

Figure L16.2: Authors

Lab 17

Figures L17.1: User:Vishalsh521, Wikimedia Commons, CC BY-SA 3.0. *http://commons.wikimedia.org/wiki/File:Katydid_india.jpg*

Figure L17.2: User:Sue in az, Wikimedia Commons, Public domain. *http://commons.wikimedia.org/wiki/File:Creosote_Larrea_tridentata.JPG*

Figure L17.3: *http://ccl.northwestern.edu/netlogo/models/BugHuntCamouflage.* BugHuntCamouflage via authors

Figure L17.4: Authors

Checkout Questions figures: A snowshoe hare with a white fur: D. Gordon E. Robertson, Wikimedia Commons, CC BY-SA 3.0. *http://commons.wikimedia.org/wiki/File:Snowshoe_Hare,_Shirleys_Bay.jpg*; A snowshoe hare with a brown fur: U.S. Fish and Wildlife Service, Wikimedia Commons, Public domain. *http://commons.wikimedia.org/wiki/File:Snowshoe_hare_eating_grass.jpg.*

Lab 18

Figure 18.1: Authors

Figure 18.2: Authors

Figure 18.3: Authors

Figure 18.4: Authors

Figures L18.1: User:TUBS, Wikimedia Commons, CC BY-SA 3.0, GFDL 1.2. *http://commons.wikimedia. org/wiki/File:Galapagos_Islands_in_South_America_ (-mini_map_-rivers).svg*

Figure L18.2: Mike Weston, Wikimedia Commons, CC BY 2.0. *http://commons.wikimedia.org/wiki/ File:Daphne_Major.jpg*

Figure L18.3: User:Charlesjsharp, Wikimedia Commons, CC BY-SA 3.0. *http://commons. wikimedia.org/wiki/File:Female_Galápagos_ medium_ground_finch.jpg*

Figures L18.4: Forest & Kim Starr, Wikimedia Commons, CC BY 3.0. *http://commons.wikimedia. org/wiki/File:Starr_060228-6323_Tribulus_cistoides. jpg*

Figure L18.5: User:6th Happiness, Wikimedia Commons, CC BY-SA 3.0. *http://commons. wikimedia.org/wiki/File:6H-diGangi-Purslane-Seed-Pods.jpg*

Figure L18.6: Authors

Checkout Questions figure: Roger W. Barbour, United States Fish and Wildlife Service, Wikimedia Commons, Public domain. *http://commons. wikimedia.org/wiki/File:Peromyscus_polionotus_ oldfield_mouse.jpg*

Lab 19

Figure 19.1: Authors

Figure 19.2: Authors

Figure 19.3: Authors

Figure 19.4: Authors

Figure L19.1: Authors

Figure L19.2: Authors

Figure L19.3: Authors

Figure L19.4: Authors

Figure L19.5: Ryan Somma, Wikimedia Commons, CC BY-SA 2.0. *http://commons.wikimedia.org/wiki/ File:Seymouria.jpg*

Figure L19.6: Authors

Checkout Questions figure: Authors

Lab 20

Figure L20.1: John Gould, Wikimedia Commons, Public domain. *http://commons.wikimedia.org/wiki/ File:Darwin%27s_finches_by_Gould.jpg*

Figure L20.2: Волков Владислав Петрович, Wikimedia Commons, Public domain. *http:// commons.wikimedia.org/wiki/File:Homology_ vertebrates.svg*

Figure L20.3: Authors

INDEX

Page numbers printed in **boldface** *type refer to figures or tables.*

National Science Teachers Association